"YOU NEED TO [...]
TELLIGENT LIFE [...]
RED SUN BETELGEUSE. BUT FIRST YOU NEED TO
KNOW HOW TO *PRONOUNCE* BETELGEUSE. WHO
YOU GONNA CALL? ISAAC ASIMOV, OF COURSE."
—*Kirkus Reviews*

- *What are the hidden forces that bind the universe together?*
- *How does artificial intelligence compare with the human mind?*
- *What are the properties of chaos?*
- *How will the discovery of a subatomic particle tell us about the origins of the universe?*

The answers to these questions and others are revealed with the intellectual drama and drive, touches of humor, surprise, and even suspense that bring readers back to Asimov time and again.

"*Engrossing . . . Isaac Asimov makes physics interesting. Ditto biology, chemistry, astronomy and other fields of science.*"
—*Anniston Star*

"The most amazing thing about Isaac Asimov is not that he is so prolific, but that his writing is so consistently engaging."
—American Library Association

ISAAC ASIMOV is the world's most prolific and popular writer of science fiction and nonfiction, including such works as *The Chemicals of Life, The Human Brain*, and the *Understanding Physics* series (all available in Mentor editions). He has authored more than 300 books, and his most recent bestseller is the novel *The Robots of Dawn*.

THE
SUBATOMIC
MONSTER

ISAAC ASIMOV

A MENTOR BOOK

NEW AMERICAN LIBRARY

NEW YORK AND SCARBOROUGH, ONTARIO

The following essays in this volume are reprinted from *The Magazine of Fantasy and Science Fiction*, having appeared in the indicated issues:

The Properties of Chaos (June 1983)
Green, Green, Green Is the Color . . . (July 1983)
What Truck? (August 1983)
Where All the Sky Is Sunshine (September 1983)
Updating the Satellites (October 1983)
More Thinking About Thinking (November 1983)
Arm of the Giant (December 1983)
The World of the Red Sun (January, 1984)
The Subatomic Monster (February 1984)
Love Makes the World Go Round! (March 1984)
E Pluribus Unum (April 1984)
Up We Go (May 1984)
The Two Masses (June 1984)
The Victorious General (July 1984)
Coming Full Circle (August 1984)
The Different Years of Time (September 1984)
The Different Years of the Universe (October 1984)

MENTOR TRADEMARK REG. U.S. PAT. OFF. AND FOREIGN COUNTRIES
REGISTERED TRADEMARK—MARCA REGISTRADA
HECHO EN CHICAGO, U.S.A.

SIGNET, SIGNET CLASSIC, MENTOR, ONYX, PLUME, MERIDIAN and NAL BOOKS
are published *in the United States* by New American Library,
1633 Broadway, New York, New York 10019,
in Canada by The New American Library of Canada Limited,
81 Mack Avenue, Scarborough, Ontario M1L 1M8

First Mentor Printing, November, 1986

1 2 3 4 5 6 7 8 9

PRINTED IN THE UNITED STATES OF AMERICA

CONTENTS

CONTENTS

INTRODUCTION

With over three hundred published books to my credit, I have come to accept the fact that I am a "prolific writer." It is what I am invariably called.

I'm not sure that, given my choice, that is quite the inevitable two-word combination I would wish to be blessed with. I suppose that it would be nicer if I were routinely called, let us say, "gifted writer" or "terrific writer" or even "writing genius." Unfortunately, it looks as though it might be a long, cold wait for any of those tie-ins to turn up, so I will accept "prolific writer."

As I think of it, there are advantages to being one. For one thing, if you are a prolific writer it is inevitable that writing is easy for you. You cannot, at one and the same time, suffer agonies as you squeeze out word by bitter word in drops of wormwood and be prolific too. There aren't enough minutes in the hour for that and there certainly isn't enough endurance in the human soul.

As a matter of fact, writing *is* easy for me, and enjoyable, too. Incredibly enjoyable, so I'm a fortunate man.

INTRODUCTION

What's more, if you're a prolific writer, you've got to be able to write quickly. You have no choice. You can write slowly, if you wish to. Or you can turn out twenty-one books in one year (including some relatively small ones, of course)—as I did in 1983. What you *cannot* do is write twenty-one books in one year and do it *slowly*.

Well, I do write quickly—very quickly. Not to beat around the bush, I write as quickly as I type and, on the word processor, I do a hundred words a minute (if we don't count time lost in correcting typos). Writing quickly is the biggest advantage I have.

Which brings me to the seamy side of being prolific, for it has its disadvantages, too—and its biggest disadvantage is that I write quickly. Yes, it's a disadvantage too.

As you can see by this book you're holding (or at least you will after you finish reading it), I write, with seeming authority, on a wide variety of subjects. And if you read the more than two dozen other books of science essays (to say nothing of the books I've written on other subjects from commentaries on the Bible to books on humor), the range would seem even more astonishing.

Well, I'm going to get away with this I-know-everything aura that I somehow build up about myself, it is absolutely necessary that I refrain from making silly mistakes. And I would, too, were it not for my very inconvenient speed of writing. I might accidentally say something ridiculous and then, before I have a chance to look at it and say, "Hey, that's wrong!" I'm in the next paragraph with my mind on something else altogether.

In the chapter "Arm of the Giant" in this book, I carefully calculated the size of the star Betelgeuse by trigonometry, and then must have confused the radius and the diameter, and ended up by making the star precisely twice the diameter it ought to be.

INTRODUCTION

I sent away "Arm of the Giant" to *The Magazine of Fantasy and Science Fiction*, which prints these articles first-time-round, and then a month later wrote a continuation of the discussion, the chapter entitled "The World of the Red Sun."

I needed the size of Betelgeuse again, and was too lazy to look up the earlier essay. I just recalculated it, and this time did not make the mistake and got the right figure. Did I notice that the two figures in the two articles were not identical? Of course not.

How did I find out (since I obviously know it now)? Easy! As soon as the first essay appeared in the magazine, Mr. Jogn (*sic*) Fortier, a self-described "devoted and addicted Reader" took typewriter in hand to point out my mistake. He didn't even use trigonometry for the purpose, just ordinary arithmetical computations. (I could have done the same if I had only been smart enough.)

What's more, he pointed out something even more ridiculous in the very same article. It came about this way—

I wanted to make mention of Jupiter's apparent size, its "angular diameter" as seen in Earth's sky. It was of no importance to the article really—it was a mere aside, dragged in by the scruff of its neck. I looked up Jupiter's angular diameter, and the maximum value, when it is closest to Earth, is 50 seconds of arc.

Correct! Very good! Except that somehow between the time my eyes left the page of the reference book and the time my eyes hit my typewriter, a strange sea-change had transmuted the phrase in my head to "50 minutes of arc."—But I wanted seconds of arc, and I knew very well that each minute of arc equaled 60 seconds of arc, so I multiplied 50 by 60 and dutifully typed out the statement that the angular diameter of Jupiter was 3,000 seconds of arc.

That's the way it appeared in the magazine, too. And if I

had stopped racing along the typewriter just long enough to think for a fifth of a second, I would have remembered that the Moon had an angular diameter of 30 minutes of arc of 1,800 seconds of arc, and that I was therefore giving Earth's sky a Jupiter that was much larger in appearance than the Moon was.

My face turned the prettiest cherry-red when Mr. Fortier pointed that out. Of course I destroyed the misstatement at once, and the correct figure appears in this book.

Or, again, in the chapter entitled "Where All the Sky Is Sunshine" I originally made a gratuitous statement to the effect that gold, except for the artificial value placed on it by people because of its beauty and rarity, was worthless; that it had no uses worth mentioning.

At once two very dear friends of mine, Lester del Rey (who is mentioned in the introduction to the last chapter) and Jay Kay Klein, wrote me long letters, listing all kinds of uses that gold would have if it were only plentiful and cheap. For this book, I dropped the offending sentence as though it had turned red-hot in my hands. Which, in a way, it had.

And there you have it. I get away with writing as quickly as I do because I have eagle-eyed readers who check every remark I make and report every error and infelicity to me at once so that I may correct it and learn from my mistake.

What would I do without them? May I take this opportunity to thank everyone—*everyone*—who has ever sent in a letter of correction and helped educate me? Let me say, too, that all the letters of this sort that I have ever received have, without exception, been couched in the most pleasant and best-natured tones. I thank you all humbly for that, too.

PART I
PHYSICS

1

THE SUBATOMIC MONSTER

Every once in a while I am told I have "missed my calling." This, of course, is invariably said in good-natured mockery, usually when I have given a funny talk or sung a comic song. The idea, then, is that I ought to have been a stand-up comedian, or a singer, perhaps.

I can't very well let the statement go unchallenged and, by long experience, I have discovered that the most effective reply to the yell, "You missed your calling, Isaac," is—

"I know, my friend, but who wants a gray-haired old stud?"

Except that nothing is foolproof. I have used that squelch at least fifty times with the greatest success, and then a few days ago when I tried it, there came back the instant reply—

"A gray-haired old nymphomaniac!"

And with that, the tables were very neatly turned, and I had to wait quite a while for the laughter to stop. (My own included.)

But I haven't missed my calling, really, and everyone

15

PHYSICS

knows it. My calling is that of a writer, and that's what I am.
In particular, my calling is that of an explainer, and *that's*
what I am, too. So if you don't mind, I will now go about my
calling.

For instance, how do you measure energy?

Well, work involves the expenditure of energy, and it is
therefore energy in action, so to speak. One way of defining
work is to say that it involves the overcoming of a resistance
over some particular distance. You overcome the resistance
by exerting a force.

For instance, Earth's gravitational pull tends to keep an
object on the ground in place. To lift it, you have to exert a
force to overcome the gravitational resistance.

The greater the weight of the object being lifted, the greater
the force you must exert and the more work you do. The
greater the distance through which you lift the weight, the
more work you do. Work, then (and energy expended), is
force times distance.

If you lift a weight of 1 pound through a distance of 1 foot
you have done 1 "foot-pound" of work. (Notice that you put
the distance first in this unit of work. There's no reason you
couldn't put the weight first and call it 1 "pound-foot" but
no one does that, and in all languages and cultures the
explanation that "no one does that" is the most unanswerable
stopper there is.)

If, then, you should happen to weigh 150 pounds and
should climb a flight of stairs that raises you 8 feet you have
done 150 × 8, or 1200 foot-pounds of work. Since I have
observed that there are often 13 steps to a flight of stairs, the
work done by a 150-pounder in moving up one step is
$1200/13$, or 92.3 foot-pounds.

But "feet" and "pounds" are units in the common system
which, to scientists, is beneath contempt. The metric system

16

is in universal use outside the United States, and is used by scientists even inside the United States. The unit of distance in the metric system is the meter, which is equal to 3.281 feet; the kilogram, which is equal to 2.2046 pounds, is used for weight.

A unit of energy in the metric system would therefore be 1 "kilogram-meter" (here the weight is first, and you don't say "meter-kilogram" because—all together, now—"no one does that"). One kilogram-meter is equal to 2.2046 pounds times 3.281 feet, or 7.233 foot-pounds. Therefore for a 150-pound person to move up one step on a flight of stairs is to do 12.76 kilogram-meters of work.

Using weight as part of a unit of work is not ideal. It is not wrong to do so, for weight is a force, but that is precisely the problem. The units popularly used for weight (pounds, or kilograms) are not, strictly speaking, units of force, but are units of mass. The confusion arises because weight has been understood since prehistoric times, whereas the concept of mass was first made clear by Isaac Newton, and mass is so similar to weight under ordinary circumstances that even scientists fell into the trap of using the long-established units of weight for mass as well, thus creating the confusion.

If we forget weight and deal only with mass, then the definition of force (which arises from Newton's second law of motion) is that of mass multiplied by acceleration. Suppose we imagine a force capable of accelerating a mass of 1 kilogram by an amount equal to 1 meter per second per second. That force is equal to 1 kilogram-meter per second per second, or (using abbreviations), 1 kg-m/sec^2. For the sake of brevity, 1 kg-m/sec^2 is called "1 newton" in honor of the great scientist. As it happens, the force required to lift a weight of 1 kilogram is 9.8 newtons. Conversely, 1 newton is the force required to lift a weight of 0.102 kilograms.

Since work is force times distance, the unit of work would be 1 newton of force expended over a distance of 1 meter. This would be 1 newton-meter. The newton-meter is usually referred to as the "joule," after the English physicist James Prescott Joule, who did important work on energy. The unit of work then is 1 joule, and since the newton is equal to a weight of 0.102 kilograms, 1 joule is equal to 0.102 kilogram-meters. Consequently, lifting 150 pounds up one step of a flight of stairs involves an amount of work equal to 125 joules.

As you see, the joule is a good unit of energy for everyday life, since an ordinary action involves a small number that is easily handled.

Suppose, however, that you wanted to deal with much smaller bits of work or energy. You would be involved with tiny fractions of a joule. It might then be helpful to have a smaller unit.

Instead of a force that imparts to 1 kilogram an acceleration of 1 meter per second per second, imagine a force that imparts to 1 gram an acceleration of 1 centimeter per second per second. Now you have a force of 1 gram-centimeter per second per second, or 1 g-cm/sec^2, which can be defined as "1 dyne" (the first syllable of a Greek word meaning "power").

Since a gram is $\frac{1}{1000}$ of a kilogram, and a centimeter is $\frac{1}{100}$ of a meter, a force of 1 dyne produces $\frac{1}{100}$ the acceleration in $\frac{1}{1000}$ the mass as compared with a force of 1 newton. Consequently 1 dyne is equal to $\frac{1}{1000} \times \frac{1}{1000}$, or $\frac{1}{100,000}$ newtons. That is the same as saying that 1 newton = 100,000 dynes.

If we suppose that 1 dyne is expended through a distance of 1 centimeter, that gives us as a unit of work "1 dyne-centimeter", or "1 erg" (the first syllable of a Greek word meaning "work"). Since a joule is the result of a newton expended over a distance of 1 meter, while an erg is the result of a dyne (1/100,000 of a newton) expended over a distance of 1 centimeter (1/100 of a meter), 1 erg is equal to 1/100,000 × 1/100, or 1/10,000,000 of a joule. That's the same as saying that 1 joule = 10,000,000 ergs.

A 150-pound person going up one step of a flight of stairs does 13,000,000 ergs of work. This is a number that is very inconvenient for everyday life, but is very handy for scientists working with small amounts of energy.

However, even the erg is by far too large a unit when it comes to dealing with individual atoms and subatomic particles. For such things, we need a smaller unit still.

Thus, instead of using a mass of a kilogram or a gram, let us use the smallest mass definitely known to exist. This is the mass of the electron, which is 0.00000000000000000000000-0000091095 grams, or 9.1095×10^{-28} grams. To avoid all those zeros, we can take the mass of an electron to be equal to "1 electron."

An electron carries an electric charge, and it therefore undergoes an acceleration in an electric field. That property of the electric field that induces an acceleration is its voltage, so we can suppose that an electron is given an acceleration produced by 1 volt.*

Given the mass and charge of the electron, the work done on it when it is exposed to the acceleration produced by 1 volt is "1 electron-volt." In abbreviated form, this is "1 ev."

This is a very tiny unit of work indeed. In fact, 1 electron-volt is equal to only a little over a trillionth of an erg. To be more precise, 1 electron-volt = 0.0000000000016 ergs, or 1.6×10^{12} ergs. (Remember, by the way, that all units of work serve as units of energy as well.)

As it happens, mass is a form of energy—a very concen-

---

*I am resisting the impulse to explain the various electrical units. That is for another essay at another time.

trated form. Mass can therefore be expressed in units of energy, but mass is such concentrated energy that ordinary units of energy are inconvenient to use in connection with ordinary masses.

Take a mass of 1 gram, for instance. This isn't much. It is only the mass of a half-grown hummingbird. The energy equivalent of that mass, by Albert Einstein's famous equation, is $e = mc^2$, where e is the energy, m is the mass, and c is the speed of light. We are taking the mass as 1 gram, and the speed of light is 29,980,000,000 centimeters per second (a combination that will give us the energy equivalent in ergs). The energy content of 1 gram of mass is, then, $1 \times 29,980,000,000 \times 29,980,000,000 = 898,800,000,000,000,000,000,000$, or 8.988×10^{20} ergs. You'll admit it's a lot easier to talk about 1 gram than about nearly a sextillion ergs.

When we get down to the electron, however, things are reversed. The tiny mass of an electron, 9.1095×10^{-28} grams, when multiplied by the energy equivalent of 1 gram, which is 8.988×10^{20} ergs, yields the energy equivalent of an electron's mass as 8.1876×10^{-7} ergs. In other words, the energy equivalent of an electron's mass is a little less than a millionth of an erg, which is hard to handle.

If we convert that energy equivalent, however, into electron-volts, which are far tinier than ergs, it turns out that the energy equivalent of an electron's mass is equal to about 511,000 electron-volts.

Of course, 511,000 might still be considered a number that is a bit too large for convenience, but 1,000 electron-volts is equal to 1 kilo-electron-volt (kev), and 1,000,000 electron-volts is equal to 1 mega-electron-volt (Mev), so you can say that the energy-equivalent of an electron's mass is about half a Mev.

The electron (and its opposite number, the positron) have, as I said earlier, the smallest masses of any object that we definitely know of as having mass at all. It may even be that there cannot be any smaller mass that is still greater than zero. There is some possibility that the various neutrinos may have still smaller masses—masses as little as 40 electron-volts—but that is as yet far from established.

What about more massive particles?

Electrons make up the outer regions of the atoms, but protons and neutrons make up the atomic nuclei, and the protons and neutrons are considerably more massive than the electrons. A proton has the energy equivalent of 938,200,000 electron-volts, or 938.2 Mev, and is thus 1,836 times as massive as the electron. The neutron has an energy equivalent of 939,500,000 electron-volts, or 939.5 Mev, and is thus 1,838.5 times as massive as the electron and 1.0014 times as massive as the proton.

An energy of 1,000,000,000 electron-volts is 1 giga-electron-volt (1 Gev), so we can say that the proton and neutron are each very nearly 1 Gev in energy-equivalence.

There are subatomic particles more massive than the proton and neutron. For instance, the W particle (something I may take up in a future essay) has recently been discovered, and it is roughly 80 times as massive as a proton, so that its energy equivalence is about 80 Gev, or 80,000,000,000 electron-volts. The nuclei of the most massive elements now have energy equivalences in the neighborhood of 250 Gev, which is more than three times larger still, but those nuclei are conglomerates of over 250 subatomic particles.

If we want a real subatomic monster, however, we will have to have a digression first.

Electricity and magnetism are closely related; in fact, they are inseparable. Everything that has an electric field has a magnetic field, and vice-versa. In fact, scientists commonly speak of an electromagnetic field, rather than of either an electric or magnetic field separately. They speak of light as an electromagnetic radiation, and of the electromagnetic interaction as one of the four fundamental interactions of nature.

Naturally, then, it is not surprising that electricity and magnetism, when viewed separately, show many similarities. Thus a magnet has two poles, showing opposite extremes, so to speak, of magnetic properties. We call them "north pole" and "south pole." There is an attraction between north poles and south poles, and a repulsion between two north poles or between two south poles.

Similarly, an electrical system has two opposite extremes, which we call "positive charge" and "negative charge." There is an attraction between a positive and negative charge, and a repulsion between two positive charges or between two negative charges.

In each case the attraction and repulsion are of equal intensities, and both the attraction and repulsion fall off in inverse proportion to the square of the distance.

There remains, however, an enormous difference of one kind.

Suppose you have a rod of insulating material in which, in one way or another, you have produced at one end a negative charge, and at the other end a positive charge. If, then, you break the rod in the middle, one half is entirely negative in charge, the other half is entirely positive. What is more, there are subatomic particles, like the electron, that carry a negative charge only, and others, like the protons, that carry a positive charge only.

Suppose, though, you have a long magnet, with a north pole at one end and a south pole at the other. If you break that in the middle, is one half entirely north pole and the other half entirely south pole?

No! If you break a magnet in two, the north pole half at once develops a south pole at the break, while the south pole half develops a north pole at the break. Nothing you can do will produce any object possessing only one magnetic pole; both are always present. Even subatomic particles that have an electric charge and, therefore, an associated magnetic field have both a north pole and a south pole.

Nor do there seem to be particular subatomic particles that carry only north poles or only south poles, though there are countless subatomic particles that carry only positive charges or only negative charges. There seems to be no such thing, in other words, as a "magnetic monopole."

About 1870, when the Scottish physicist James Clerk Maxwell first worked out the mathematical relationships that described the electro-magnetic field as a unified phenomenon, he presented the world with four concise equations that seemed totally sufficient for the purpose for which they were designed. If there had existed magnetic monopoles, the four equations would have been beautifully symmetrical so that electricity and magnetism would have represented a kind of mirror image of each other, As it was, however, Maxwell assumed that magnetic poles always existed in pairs, while electric charge did not, and that, perforce, introduced an asymmetry.

Scientists dislike asymmetries, for they offend the aesthetic sense and interfere with simplicity (the be-all and end-all of perfect science), so there has been a constant feeling that the monopole *ought* to exist; that its nonexistence represents a flaw in cosmic design.

After the electron was discovered, there eventually came the realization that electric charge is quantized—that is, that all electric charges are exact multiples of some fundamental smallest value.

Thus all electrons have an identical negative charge and all protons an identical positive charge, and the two kinds of charge are exactly equal to each other in size. All other known charged objects have an electric charge that is exactly equal to that of the electron, or of the proton, or that is an exact multiple of one or the other.

(Quarks are thought to have charges equal to $\frac{1}{3}$ and $\frac{2}{3}$ that of the electron or proton, but quarks have never been isolated—and even if they were, that would merely make the fundamental smallest value a third of what it had been thought to be. The principle of quantization would remain.)

Why should electric charge be quantized? Why couldn't it exist in any uneven value, just as mass does? After all, the mass of a proton is a thoroughly uneven multiple of the mass of an electron, so why shouldn't this be true of charge as well?

In 1931 the English physicist Paul A. M. Dirac tackled the matter mathematically and came to the decision that this quantization of charge would be a logical necessity if magnetic monopoles existed. In fact, even if only one monopole existed anywhere in the universe, quantization of charge would be a necessity.

It is tempting to argue the reverse, of course: that since electric charge *is* quantized, magnetic monopoles must exist somewhere. It made increasing sense to search for them.

But where and how do you find them even if they exist? Physicists didn't know and, what was worse, they weren't sure what the properties of these monopoles might be. It seemed a natural assumption to suppose they were fairly massive particles, because if so they wouldn't be very common and couldn't be easily brought into existence in the laboratory—and that would explain why no one had happened to stumble across one accidentally.

There was no theoretical guideline farther than that until the 1970s, when people were working out various grand unified theories designed to combine the weak, strong, and electromagnetic interactions all under a single set of equations (see "And After Many a Summer Dies the Proton" in *Counting the Eons*, Doubleday, 1983).

In 1974 a Dutch physicist, Gerard't Hooft, and a Soviet physicist, Alexander Polyakov, independently showed that it could be reasoned from the grand unified theories that magnetic monopoles *must* exist and that they are not merely massive, they are *monsters*.

Although a monopole would be even smaller than a proton, packed into its tininess might be a mass of anywhere from ten quadrillion to ten quintillion times that of the proton. If it is at the upper end of this range, a monopole would have an energy equivalence of 10,000,000,000,000,000,000,000,000,000,000,000,000 electron-volts (10^{28} ev).

How much would that be in mass? It turns out that a magnetic monopole might have a mass of as much as 1.8×10^{-9} grams. That is equal to the mass of 20 human spermatozoa all packed into a single subatomic particle.

How can such subatomic monsters be formed? There's no way human beings can pack that much energy into a subatomic volume of space, either now or in the foreseeable future. In fact, there's no natural process taking place anywhere in the universe right now (as far as we know) that could create a particle of such monstrous mass.

The only possibility is to go back to the Big Bang, when temperatures were incredibly high and energies incredibly concentrated (see "The Crucial Asymmetry" in *Counting the Eons*, Doubleday, 1983). It is estimated that the monopoles must have been formed only 10^{-34} seconds after the Big Bang. After that the universe would have been too cool and too large for the purpose.

Presumably, both north and south monopoles were formed, possibly in enormous quantities. Presumably, a great many of them annihilated one another, but a number must have survived simply because by sheer chance they did not happen to encounter others of the opposite persuasion. After monopoles had survived a certain length of time, the steady expansion of the universe made it less and less likely that collisions would take place, and that ensured their further survival. There are therefore a certain number floating around the universe today.

How many?—Well, not too many, for above a certain number the gravitational effect of these monstrous particles would have made it certain that the universe would, before now, have reached a maximum size and collapsed again under its own gravitational pull. In other words, we can calculate a maximum monopole-density in the universe by simply recognizing the fact that we ourselves exist.

Yet even though few in number a monopole ought, every once in a while, move into the vicinity of a recording device. How would it then be detected?

Scientists had earlier expected that monopoles would be moving at nearly the speed of light, as cosmic ray particles do; and that like cosmic ray particles, monopoles would smash up other particles in their path and produce a shower of secondary radiation that could be easily detected and from which the monopole itself could be identified.

Now that the monopole is thought to be of monstrous mass, things have changed. Such huge monopoles couldn't accumulate enough energy to move very rapidly, and it is estimated that they would be traveling at a speed of a couple of hundred kilometers per second—less than a thousandth the speed of light. At such slow speeds monopoles would simply slip quietly past and through matter, without leaving any signs to speak of. It may be that which accounts for the failure to have detected monopoles hitherto.

Well, then, what is to be done?

A physicist at Stanford University, Blas Cabrera, had an idea. A magnet pushing through a coil of wire will send a surge of electric current through that coil. (This has been known for over a century and a half.) Why not set up such a coil, then, and wait? Perhaps a magnetic monopole will just happen to pass through the coil and signal its passage by an electric current. Cabrera calculated the chances of that happening on the basis of the top monopole density that is possible in view of the fact that the universe exists and decided that one such event might happen every six months on the average.

Cabrera therefore set up a coil of the metal, niobium, and kept it at a temperature near absolute zero. Under these conditions, niobium is superconducting and has zero resistance to an electric current. This means that if anything starts a current flowing in it, that current will flow indefinitely. A monopole passing through the coil won't just give an instantaneous surge of current, but a current that continues and continues.

Naturally, a current could be started by any old magnetic field that happens to be around—the Earth's own magnetic field, those set up by any of a number of technical devices in the vicinity, even by stray bits of metal that happen to be moving because they are in someone's pocket.

Cabrera therefore placed the coil inside a superconducting lead balloon, which was inside a second superconducting lead balloon. Ordinary magnetic fields would not penetrate the superconducting lead, but a magnetic monopole would.

He waited for four months and nothing happened. The current level marked out on a moving roll of paper stayed near zero throughout. This was, in itself, good. It showed he had successfully excluded stray magnetic fields.

Then at 1:53 P.M. on February 14, 1982, there came a sudden flow of electricity—and in just about exactly the amount one would expect if a magnetic monopole had passed through.

Cabrera checked every possible event he could think of that

would have started the current *without* the help of a monopole, and could find nothing. The monopole seemed the only alternative.

Has the elusive monopole been detected, then? If so, it is a remarkable feat and it strongly supports the grand unified theory.

The trouble is, though, that there has been no repetition of that single event, and it is hard to base anything on just one happening.

Then, too, Cabrera's estimate of the number of monopoles floating around was based on the fact that the universe is still expanding. Some people think that a stronger constraint arises out of the possibility that monopoles floating about the galaxy would wipe out the general galactic magnetic field. Since the galactic magnetic field still exists (though it is very weak), that could set a far lower maximum value on monopole density—as little, perhaps, as $1/10,000$ of Cabrera's figure.

If that were so, one would expect a monopole to pass through his coil once every 5,000 years, on the average. And if *that* were so, to have one pass through after only four months of waiting is asking an awful lot, and it becomes difficult to believe that it was a monopole.

There is only one thing to do, and physicists are doing it. They are continuing the search. Cabrera is building a larger and better version of his device that will increase his chances of detecting a monopole fiftyfold. Other physicists are figuring out other ways to go about the detection.

Over the next few years the search for the monopole is going to increase enormously in intensity, for the stakes are high. The definite detection will give us an indication of the properties of the subatomic monster and its numbers. From that, we may learn things about the beginning of the universe, to say nothing of its present and future, that we might not otherwise ever find out.

And, of course, there's a certain Nobel Prize waiting for somebody.

2

E PLURIBUS UNUM

My dear wife, Janet, is a writer quite on her own, having made a few sales before ever she met me. At present, she has published two novels *(The Second Experiment* and *The Last Immortal)* under her maiden name, J. O. Jeppson, and has collaborated with me on an anthology of humorous science fiction (including verse and cartoons) entitled *Laughing Space*. All three books were published by Houghton Mifflin. In addition, a book of her short stories has been published by Doubleday.

Best of all, she has published a lighthearted science fiction juvenile entitled *Norby, the Mixed-up Robot* (Walker, 1983), in collaboration with me, and the authorship even acknowledges our marriage. It's by ''Janet and Isaac Asimov.'' It is the first of a series, and the second, *Norby's Other Secret*, was published in 1984. It's pleasant to be unified in print this way.

In fact, unification is pleasant in many fields. Americans are surely glad that thirteen independent states decided to unite in a single federal government. It is that which has made *E pluribus unum* (Latin for "Out of many, one") a phrase that is so associated with the United States. And scientists like to unify, too, taking pleasure in showing that events that may seem totally distinct are actually different aspects of a single phenomenon.

Let's begin with "action at a distance."

Ordinarily, if you want to achieve some action such as imparting motion to an object at rest, you must make physical contact with it directly or indirectly. You might strike it with a hand or foot, or with a bat or club that you are holding. You might hold it in your hand, while making your hand move, and then release it. You might throw an object in this manner and have it hit a second object, to which it then imparts motion. In fact, you can move one object and have that motion transmitted bit by bit to many objects (as in the falling of a row of dominoes). You can even blow, making the air move and, by its impact, move something else.

Could you, however, make a distant object move without touching it, and without allowing anything you have previously touched to touch it? If so, that would be action at a distance.

For instance, suppose you are holding a billiard ball at eye level over the ground. You hold it so that it is perfectly motionless and then you suddenly let go. You have been touching it, true, but in letting go you *cease* to touch it. It is only after you cease touching it that it falls to the ground. It has been made to move without anything making physical contact with it.

Earth attracts the ball, and that is referred to as "gravitation." Gravitation seems to be an example of action at a distance.

Or consider light. If the Sun rises, or a candle is lit, a room is at once illuminated. The Sun or the candle causes the illumination without anything material seeming to intervene in the process. It, too, seems to be action at a distance. For that matter, the sensation of heat from the Sun or from the candle can be felt across a gap of space. That is another example.

Then, too, about 600 B.C. the Greek philosopher Thales (624–546 B.C.), is supposed to have studied, for the first time, a black rock that had the ability to attract iron objects at a distance. Since the rock in question came from the neighborhood of the Greek city of Magnesia on the Asia Minor coast, Thales called it *ho magnetes lithos* ("the Magnesian rock") and the effect has been called "magnetism" ever since.

Thales also discovered that if an amber rod is rubbed, it becomes capable of attracting light objects at a distance. The amber rod will attract objects a magnet won't affect, so it is a different phenomenon. Since the Greek word for amber is *elektron*, the effect has been called "electricity" ever since. Magnetism and electricity seem to represent action at a distance as well.

Finally, there is sound and smell. Ring a bell at a distance and you hear it even though there is no physical contact between the bell and yourself. Or set a steak sizzling over a flame and you will smell it at a distance.

We have seven such phenomena, then: gravitation, light, heat, magnetism, electricity, sound, and smell.

As it happens, scientists are uncomfortable with the notion of action at a distance. There are so many examples of effects that can only be brought about by some sort of contact that the few examples that seem to omit contact ring false. Perhaps there *is* contact, but in so subtle a manner that we overlook it.

Smell is the easiest of these phenomena to explain. The steak over a fire is sizzling, sputtering and smoking. Tiny particles of it are obviously being given off and are floating in the air. When they reach your nose they interact with its membranes and you interpret this as a smell. With time, this was thoroughly confirmed. Smell is a phenomenon that involves contact, and it is *not* action at a distance.

As for sound, the Greek philosopher Aristotle (384–322 B.C.), about 350 B.C., having noted that objects emitting sounds were vibrating, suggested that the vibrations struck the air immediately in its vicinity and set it to vibrating; that air set the air beyond it to vibrating and so on—like a set of invisible dominoes. Eventually the progressive vibration reached our ear and set it to vibrating so that we heard sound.

In this, as it happened, Aristotle was perfectly correct, but how could his suggestion be tested? If sound is conducted by air, then it should *not* be conducted if there is no air. If a bell is rung in a vacuum, it should emit no sound. The trouble was that neither Aristotle nor anyone else in his time, or for nearly two thousand years thereafter, could produce a vacuum and test the matter.

In 1644 the Italian physicist Evangelista Torricelli (1608–47) upended a long tube filled with mercury into a dish of mercury and found that some poured out. The weight of Earth's atmosphere only upheld thirty inches of mercury. When the mercury poured out, it left behind it between the sunken level and the closed end of the tube a gap that contained nothing, not even air—at least nothing except for some tiny traces of mercury vapor. In this way the first decent vacuum was created by human beings, but it was a small one, sealed off, and not very useful for experimentation.

Just a few years later, in 1650, the German physicist Otto von Guericke (1650–86) invented a mechanical device that, little by little, sucked air out of a container. This enabled him to form a vacuum at will. For the first time, physicists were able to experiment with vacuums.

PHYSICS

In 1657 the Irish physicist Robert Boyle (1627–91) heard of Guericke's air pump, and had his assistant, Robert Hooke (1635–1703), devise a better one. In no time at all he showed that a bell that was set to ringing in a glass container that had been evacuated made no sound. As soon as air was allowed to enter the container, the bell sounded. Aristotle was right, and sound, like smell, did *not* represent action at a distance.

(Nevertheless, over three and a quarter centuries later, movie-makers still have spaceships move through space with a whoosh, and explode with a crash. I suppose that either movie-makers are ignorant, or, more likely, they assume the American public is and feel they have a divine right to protect and preserve that ignorance.)

The question is, then, what phenomena will make themselves felt across a vacuum. The fact that air pressure will only support a column of mercury 30 inches high means that air can only extend a few miles above Earth's surface. Beyond a height of ten miles, only relatively thin wisps of air remain. This means that the 93,000,000-mile gap between the Sun and the Earth is virtually nothing but vacuum, and yet we feel the Sun's heat and see its light, while the Earth responds to the Sun's gravitational pull by circling it endlessly. Furthermore, it was as easy to show that a magnet or an electrified object exerted their effects across a vacuum as it was to show that a ringing bell did not.

This leaves us with five phenomena that might represent action at a distance: light, heat, gravitation, magnetism, and electricity.

Nevertheless, scientists were still not anxious to accept action at a distance. The English scientist Isaac Newton (1642–1727) suggested that light consisted of a spray of very fine particles moving in rigidly straight lines. The light source would emit the particles and the eyes would absorb them; in between, the light might be reflected from something and the eyes would see that something by the light it reflected. Since the particles touched the object and then the eye, it was not action at a distance but action by contact.

This particle theory of light explained a number of things, such as the fact that opaque objects cast sharp shadows. It left some puzzles, however. Why should light passing through a prism split into a rainbow of colors? Why should the particles of red light be less refracted than those of violet light? There were explanations, but they weren't entirely convincing.

In 1803 the English scientist Thomas Young (1773–1829) conducted experiments that showed that light consisted of waves (see "Read Out Your Good Book in Verse" in *X Stands for Unknown,* Doubleday, 1984). The waves were of different lengths, those for red light twice as long as those for violet light, and the difference in refraction was easily explained in this way. The reason for sharp shadows (water waves and sound waves do not cast them) is that the wavelengths of light are so tiny. Even so, the shadows are not, in actual fact, perfectly sharp. There is a little fuzziness ("diffraction") and that could be demonstrated.

Light waves put physicists back to action at a distance with a vengeance. One could say that the waves traveled across a vacuum, but how? Water waves are propagated through the motion of surface water molecules at right angles to the direction of propagation (transverse waves). Sound waves are propagated through the motion of air molecules backward and forward in the direction of propagation (longitudinal waves). But when light waves travel across a vacuum, there is no material of any sort to move either up and down or back and forth. How, then, does the propagation take place?

The only conclusion scientists could come to was that a vacuum did *not* contain nothing; that it *did* contain something that would wave up and down (for light waves were discovered to be transverse, like water waves). They therefore postulated the existence of "ether," a word borrowed from Aristotle. It was a substance so fine and subtle that it could not be detected by the gross methods of science, but could only be inferred from the behavior of light. It permeated all of space and matter, reducing anything that seemed to be action at a distance, to action by contact—etheric contact.

(Ether was eventually found to be an unnecessary concept, but that's another story. For convenience's sake, I will temporarily speak of the various effects that can make themselves felt across a vacuum as "etheric phenomena.")

There are then the five etheric phenomena I listed a while ago, but might there not be more that would eventually be discovered, as electricity and magnetism had once been discovered by Thales? Or, in reverse, might there not be fewer? Might some etheric phenomena that seemed distinct actually prove identical when viewed in a more fundamental way?

In 1800, for instance, the German-British astronomer William Herschel (1738–1822) discovered infrared radiation— radiation beyond the red end of the spectrum. The infrared so strongly affected a thermometer that Herschel thought, at first, that that invisible region of the spectrum consisted of "heat rays."

It was not long, however, before the wave theory of light was established and it was understood that there was a much wider stretch of wavelength than that which the human eye was equipped to detect (see "Four Hundred Octaves" in *X Stands for Unknown*, Doubleday, 1984).

Heat came to be understood better, too. It could be transmitted by conduction through solid matter, or by convection along moving currents of liquid or gas. This is action by means of atoms or molecules in contact. When heat makes itself felt across a vacuum, however, so that it is an etheric phenomenon, it does so by the radiation of light waves, particularly in the infrared. These radiations are not in themselves heat, but are only perceived as such when they are absorbed by matter and the energy so absorbed sets the constituent atoms and molecules of that matter to moving or vibrating more rapidly.

Therefore we can expand the concept of light to signify the entire spectrum of lightlike waves, whether detectable by eye or not, and it can also include heat in its radiational aspect. The list of etheric phenomena is reduced to four, then: light, gravitation, magnetism, and electricity.

Is there any chance of reducing the list further? All the etheric phenomena are similar in that each originates in some source and radiates outward in all directions equally. Furthermore, the intensity of the phenomenon decreases, in each case, as the square of the distance from the source.

If you are at a given distance from a source of light and measure its intensity (the quantity of light striking a unit area), then move away until your distance is 2.512 times the original distance, the new intensity is $1/2.512^2$ or $1/6.31$ what it was at the original distance. This "inverse-square rule" can also be shown to be true of the intensity of gravitation, electricity, and magnetism.

Yet this is not, perhaps, as significant as it sounds. We might visualize each of these phenomena as a radiation moving outward at some fixed speed in all directions equally. After any particular lapse of time, the leading edge of the expanding wave occupies every point in space that is at a particular distance from the source. If you connect all these points, you will find that you have marked out the surface of a sphere. The surface of a sphere increases as the square of its radius—that is, as the square of its distance from the central point. If a fixed amount of light (or any etheric phenomenon) is spread out over the surface of an expanding sphere, then every time the surface doubles in area the amount of light available per unit area on that surface is cut in half. Since the surface area increases as the square of the distance from the source, the intensity of light (or any etheric phenomenon) decreases as the square of the distance from the source.

That means that the various phenomena might be basically different in properties and yet resemble each other in following the inverse-square law. But *are* the various etheric phenomena basically different?

Certainly they seem to be. Gravitation, electricity, and magnetism all make themselves evident as an attraction. This differentiates all three from light, which does not seem to be involved with attraction.

In the case of gravitation, attraction is the *only* effect that can be observed. With electricity and magnetism, however, there is repulsion as well as attraction. Like electric charges repel each other; so do like magnetic poles. Yet electricity and magnetism are not identical either, since the former seems capable of attracting all kinds of matter, while magnetic attraction seems largely confined to iron.

Thus in the 1780s the French physicist Charles Augustin de Coulomb (1736–1806), who had shown that both electricity and magnetism followed the inverse-square law, argued convincingly that the two might be similar in this, but were fundamentally different in essentials. That became the orthodox view.

But even as Coulomb was propounding his orthodoxy, a revolution was brewing in the study of electricity.

Until then it had been "electrostatics" that was studied, the more or less motionless electric charge on glass, sulfur, amber, and other materials that are today called nonconductors. Characteristic effects were observed when the electric content on such objects was discharged and all the charge made to flow across an air gap, for instance, to produce a spark and a crackle; or into a human body to produce a most unpleasant electric shock.

In 1791 the Italian physicist Luigi Galvani (1737–98) found that electrical effects could be produced when two different metals were in contact. In 1800 this matter was taken further by the Italian physicist Alessandro Volta (1745–1827), who made use of a series (or "battery") of two-metal contacts to produce a continuous flow of electricity. In no time at all, every physicist in Europe was studying "electrodynamics."

Yet this discovery made electricity and magnetism seem more different than ever. It was easy to produce a current of moving electric charges, but no analogous phenomenon was to be noted with magnetic poles.

A Danish physicist, Hans Christian Oersted (1777-1851), felt otherwise. Taking up the minority view, he maintained that there was a connection between electricity and magnetism. An electric current through a wire developed heat; if the wire was thin it even developed light. Might it not be, argued Oersted in 1813, that if the wire were thinner still, electricity forced through it would produce magnetic effects?

Oersted spent so much time teaching at the University of Copenhagen, however, that he had little time to experiment, and was, in any case, not a particularly gifted experimenter.

In the spring of 1820, however, he was lecturing on electricity and magnetism to a general audience and there was an experiment he wanted to try but had not had time to check before the lecture. On impulse, he tried it in the course of his lecture. He placed a thin platinum wire over a magnetic compass, running it parallel to the north-south direction of the needle, then forced a current through the wire. To Oersted's astonishment (for it was not quite the effect he expected), the compass needle jerked as the current was turned on. It wasn't much of a jerk and the audience was left unmoved, apparently, but after the lecture Oersted turned to experimentation.

He found that when the current was made to flow through the wire in one direction, the compass needle turned clockwise; when the current flowed in the other direction, it turned counterclockwise. On July 21, 1820, he published his discovery, and then dropped the matter. But he had done enough. He had established some sort of connection between electricity and magnetism, and physicists rushed to investigate the matter further with an avidity not seen again until the discovery of uranium fission over a century later.

Within days, the French physicist Dominique F. J. Arago (1786–1853) showed that a wire carrying an electric current attracted not only magnetized needles but ordinary unmagnetized iron filings, just as a straightforward magnet would. A magnetic effect, absolutely indistinguishable from that of ordinary magnets, originated in the electric current.

Before the year was over, another French physicist, André Marie Ampère (1775–1836), showed that two parallel wires that were attached to two separate batteries in such a way that current flowed through each in the same direction attracted each other. If the current flowed through in opposite directions, they repelled each other. The currents could be made to act like magnetic poles, in other words.

Ampère bent a wire into a solenoid or helix (like a bedspring) and found that the current, flowing in the same direction in each coil, produced reinforcement. The magnetic effect was stronger than it would have been in a straight wire, and the solenoid acted exactly like a bar magnet, with a north pole and a south pole.

In 1823 an English experimenter, William Sturgeon (1783–

E PLURIBUS UNUM

1850), placed eighteen turns of bare copper wire about a U-shaped iron bar without letting the wire actually touch the bar. This concentrated the magnetic effect even further, to the point where he had an "electromagnet." With the current on, Sturgeon's electromagnet could lift twenty times its own weight of iron. With the current off, it was no longer a magnet and would lift nothing.

In 1829 the American physicist Joseph Henry (1797–1878) used insulated wire and wrapped innumerable coils about an iron bar to produce a far stronger electromagnet. By 1831, he had an electromagnet of no great size that could lift over a ton of iron.

The question arose: Since electricity produces magnetism, can magnetism also produce electricity?

The English scientist Michael Faraday (1791–1867) demonstrated the answer to be affirmative. In 1831 he thrust a bar magnet into a wire solenoid to which no battery was connected. As he thrust the magnet in, there was a surge of electric current in one direction (this was easily detected with a galvanometer, which had been invented in 1820 by making use of Oersted's discovery that an electric current would deflect a magnetized needle). When he pulled the magnet out, there was a surge of electric current in the opposite direction.

Faraday then went on to construct a device whereby a copper disk was forced to turn continuously between the poles of a magnet. A continuous electric current was thus set up in the copper, and this could be drawn off. This was the first electric generator. Henry reversed matters by having an electric current turn a wheel, and this was the first electric motor.

Faraday and Henry between them thus initiated the age of electricity, and all of it stemmed from Oersted's original observation.

It was now certain that electricity and magnetism were closely related phenomena, that electricity produced magnetism and vice versa. The question now was whether each could also exist separately; whether there were any conditions under which electricity did *not* produce magnetism, and vice versa.

In 1864 the Scottish mathematician James Clerk Maxwell devised the set of four comparatively simple equations mentioned in Chapter 1. They described the nature of the interrelationships of electricity and magnetism. It quickly became apparent that Maxwell's equations held under all conditions and explained all electromagnetic behavior. Even the relativity revolution introduced by Albert Einstein (1879–1955) in the first decades of the twentieth century, a revolution that modified Newton's laws of motion and of universal gravitation, left Maxwell's equations untouched.

If Maxwell's equations were valid, then neither electrical nor magnetic effects could exist in isolation. The two were always present together, and there was just electromagnetism, within which electrical and magnetic components were directed at right angles to each other.

Furthermore, in considering the implications of his equations, Maxwell found that a changing electric field had to induce a changing magnetic field, which in turn had to induce a changing electric field, and so on. The two leapfrogged, so to speak, so that the field progressed outward in all directions in the form of a transverse wave moving at a speed of 300,000 kilometers per second. This was "electromagnetic radiation." But light is a transverse wave moving at a speed of 300,000 kilometers per second, and the conclusion was irresistible that light in all its wavelengths from gamma rays to radio waves was an electromagnetic radiation. The whole was an electromagnetic spectrum.

Light, electricity, and magnetism all melted into a single

phenomenon described by a single set of mathematical relationships—*e pluribus unum*. Now there were only two forms of action at a distance: gravitation and electromagnetism. With the vanishing of the concept of the ether, we speak of "fields"; of a "gravitational field" and an "electromagnetic field," each consisting of a source and an endlessly expanding radiation from that source, moving outward at the speed of light.

Having reduced five to two, ought we not search for some still more general set of mathematical relationships that would apply to only *one* "electromagnetogravitational field," with gravitation and electromagnetism merely two aspects of the same phenomenon?

Einstein tried for thirty years to work out such a "unified field theory" and failed. While he was trying, two new fields were discovered, each diminishing in intensity with distance so rapidly that they showed their effect only at distances comparable to the diameter of an atomic nucleus or less (hence their late discovery). They are the "strong nuclear field" and the "weak nuclear field."

In the 1870s the American physicist Steven Weinberg (1933–) and the Pakistani-British physicist Abdus Salam (1926–) independently worked out a mathematical treatment that showed the electromagnetic and weak nuclear fields to be different aspects of a single field, and this new treatment can probably be made to include the strong nuclear field as well. To this day, however, gravitation remains stubbornly outside the gate, as recalcitrant as ever.

What it amounts to, then, is that there are now two grand descriptions of the world: the theory of relativity, which deals with gravity and the macrocosm, and quantum theory, which deals with the combined electromagnetic/weak/strong field and the microcosm.

No way has yet been found to combine the two—no way, that is, to "quantize" gravitation. I can't think of any surer way of getting a Nobel Prize within a year than to accomplish that task.

3

THE TWO MASSES

I saw Albert Einstein once.

It was on April 10, 1935. I was returning from an interview at Columbia College, an interview on which my permission to enter depended. (It was disastrous, for I was a totally unimpressive fifteen-year-old, and I didn't get in.)

I stopped off in a museum to recover, for I had no illusions as to my chances after that interview, and so confused and upset was I that I've never been able to remember which museum it was. But wandering in a semi-dazed condition through the rooms, I saw Albert Einstein, and wasn't so dead to the world around me that I didn't recognize him at once.

From then on, for half an hour, I followed him patiently from room to room, looking at nothing else, merely staring at him. I wasn't alone; there were others doing the same. No one said a word, no one approached him for an autograph or for any other purpose; everyone merely stared. Einstein paid no attention whatever; I assume he was used to it.

After all, no other scientist, except for Isaac Newton, was

so revered in his own lifetime—even by other great scientists, let alone by laymen and adolescents. It is not only that his accomplishments were enormous, but that they are in some respects almost too rarefied to describe, especially in connection with what is generally accepted as his greatest accomplishment: General Relativity.

Certainly it's too rarefied for me, since I am only a biochemist (of sorts) and not a theoretical physicist, but in my self-assumed role as busybody know-it-all, I suppose I have to try anyway . . .

In 1905 Einstein had advanced his special theory of relativity (or special relativity for short), which is the more familiar part of his work. Special relativity begins with the assumption that the speed of light in a vacuum will always be measured at the same constant value regardless of the speed of the light source relative to the observer.

From that, an inescapable line of deductions tells us that the speed of light represents the limiting speed for anything in our universe—that if we observe a moving object, we will find its length in the direction of motion and the rate of time passage upon it decreased and its mass increased, as compared with what it would be if the object were at rest. These properties vary with speed in a fixed manner such that at the speed of light, length and time rate would be measured as zero while mass would be infinite. Furthermore, special relativity tells us that energy and mass are related according to the now famous equation $e = mc^2$.

Suppose, though, that the speed of light in a vacuum is *not* unchanging under all conditions. In that case, none of the deductions is valid. How, then, can we decide on this matter of the constancy of the speed of light?

To be sure, the Michelson-Morley experiment (see "The Light That Failed" in *Adding a Dimension*, Doubleday, 1964) had indicated that the speed of light did not change with the motion of the Earth—that is, that it was the same whether the light moved in the direction of Earth's revolution about the Sun or at right angles to it. One might extrapolate the general principle from that, but the Michelson-Morley experiment is capable of other interpretations. (To go to an extreme, it might indicate that the Earth wasn't moving, and that Copernicus was wrong.)

In any case, Einstein insisted later that he had not heard of the Michelson-Morley experiment at the time he conceived of special relativity and that it seemed to him that light's speed must be constant simply because he found himself involved in contradictions if that weren't so.

Actually, the best way to test Einstein's assumption would be to test whether the deductions from that assumption are to be observed in the real universe. If so, then we are driven to the conclusion that the basic assumption must be true, for we would then know of no other way of explaining the truth of the deductions. (The deductions do *not* follow from the earlier Newtonian view of the universe or from any other non-Einsteinian—or nonrelativistic—view.)

It would have been extremely difficult to test special relativity if the state of physical knowledge had been what it was in 1895, ten years before Einstein advanced his theory. The startling changes it predicted in the case of length, mass, and time with speed are perceptible only at great speeds, far beyond those encountered in ordinary life. By a stroke of fortune, however, the world of subatomic particles had opened up in the decade prior to Einstein's announcements. These particles moved at speeds of 15,000 kilometers per second and more, and at *those* speeds relativistic effects are appreciable.

It turned out that the deductions of special relativity were all there, all of them; not only qualitatively but quantitatively as well. Not only did an electron gain mass as it sped by at

nine tenths the speed of light, but its mass was multiplied by
$3\frac{1}{6}$ times, just as the theory predicted.

Special relativity has been tested an incredible number of
times in the last eight decades and it has passed every test.
The huge particle accelerators built since World War II would
not work if they didn't take into account relativistic effects in
precisely the manner required by Einstein's equations. Without the $e = mc^2$ equation, there is no explaining the energy
effects of subatomic interactions, the working of nuclear
power plants, the shining of the Sun. Consequently, no physicist who is even minimally sane doubts the validity of special
relativity.

This is not to say that special relativity necessarily represents ultimate truth. It is quite possible that a broader theory
may someday be advanced to explain everything special relativity does and more besides. On the other hand, nothing has
so far arisen that seems to require such explanation except for
the reported *apparent* separation of quasar components at
more than the speed of light, and the betting is that this is
probably an optical illusion that can be explained within the
limits of special relativity.

Then, too, even if such a broader theory is developed, it
would have to work its way down to special relativity within
the bounds of present-day experimentation, just as special
relativity works itself down to ordinary Newtonian laws of
motion, if you stick to the low speeds we use in everyday
life.

Why is special relativity called "special"? Because it deals
with the special case of constant motion. Special relativity
tells you all you need to know if you are dealing with an
object moving at constant speed and unchanging direction
with respect to yourself.

But what if the speed or direction of movement of an

object (or both) is changing with respect to you? In that case, special relativity is insufficient.

Strictly speaking, motion is never constant. There are always present forces that introduce changes in speed, direction, or both, in the case of every moving object. Consequently we might argue that special relativity is always insufficient.

So it is, but the insufficiency can be small enough to ignore. Subatomic particles moving at vast speeds over short distances don't have time to accelerate much, and special relativity can be applied.

In the universe generally, however, where stars and planets are involved, special relativity is grossly insufficient, for there we deal with large accelerations and these are invariably brought about by the existence of vast and ever-present gravitational fields.

At the subatomic level, gravitation is so excessively weak in comparison with other forces that it can be ignored. At the macroscopic level of visible objects, however, it cannot be ignored; in fact, everything *but* gravitation can be ignored.

Near Earth's surface, a falling body speeds up while a rising body slows down, and both are examples of accelerations caused entirely by progress through Earth's gravitational field. The Moon travels in an orbit about the Earth, the Earth about the Sun, the Sun about the galactic center, the galaxy about the local group center and so on, and in every case the orbital motion involves an acceleration since there is a continuing change in direction of motion. Such accelerations are also produced in response to gravitational fields.

Einstein therefore set about applying his relativistic notions to the case of motion *generally*, accelerated as well as constant—in other words, all the real motions in the universe. When worked out, this would be the general theory of relativity, or general relativity. To do this, he had, first and foremost, to consider gravitation.

PHYSICS

*   *   *

There is a puzzle about gravitation that dates back to Newton. According to Newton's mathematical formulation of the laws governing the way in which objects move, the strength of the gravitational pull depends upon mass. The Earth pulls on an object with a mass of 2 kilograms exactly twice as hard as it does on an object with a mass of only 1 kilogram. Furthermore, if Earth doubled its own mass it would pull on everything exactly twice as hard as it does now. We can, therefore, measure the mass of the Earth by measuring the intensity of its gravitational pull upon a given object; or we can measure the mass of an object by measuring the force exerted upon it by Earth.

A mass, so determined, is "gravitational mass."

Newton, however, also worked out the laws of motion and maintained that any force exerted upon an object causes that object to undergo an acceleration. The amount of acceleration is in inverse proportion to the mass of the object. In other words, if the same force is exerted on two objects, one which has a mass of 2 kilograms and the other of 1 kilogram, the 2-kilogram object will be accelerated to exactly half the extent of the 1-kilogram object.

The resistance to acceleration is called inertia, and we can say that the larger the mass of an object, the larger its inertia—that is, the less it will accelerate under the push of a given force. We can therefore measure the mass of a body by measuring its inertia—that is, by measuring the acceleration produced upon it by a given force.

A mass so determined is "inertial mass."

All masses that have ever been determined have been measured either through gravitational effects or inertial effects. Either way is taken as valid and they are treated as interchangeable, even though the two masses have no *apparent* connection. Might there not, after all, be some objects, made

THE TWO MASSES

of certain materials or held under certain conditions, that would show an intense gravitational field but very little inertia, or vice versa? Why not?

Yet whenever one measures the mass of a body gravitationally, then measures the mass of the same body inertially, the two measurements come out to be equal. Yet that may only be appearance. There may be small differences, too small to be noted ordinarily.

In 1909 an important experiment in this connection was performed by a Hungarian physicist, Roland, Baron von Eötvös (the name is pronounced "ut'vush").

What he did was to suspend a horizontal bar from a delicate fiber. At one end of the bar was a ball of one material, and at the other end a ball of another material. The Sun pulls on both balls and forces an acceleration on each. If the balls are of different mass, say 2 kilograms and 1 kilogram, then the 2-kilogram mass is attracted twice as strongly as the 1-kilogram mass, and you might expect it to be accelerated twice as strongly. However, the 2-kilogram mass has twice the inertia of the 1-kilogram mass. For that reason, the 2-kilogram mass accelerates only half as much per kilogram and ends up being made to accelerate only as strongly as the 1-kilogram mass is.

If inertial mass and gravitational mass are *exactly* equal, then the two balls are made to accelerate *exactly* equally, and the horizontal bar may be pulled toward the Sun by an immeasurable amount, but it does not rotate. If inertial mass and gravitational mass are not quite equal, then one ball will accelerate a bit more than the other and the bar will experience a slight turning force. This will twist the fiber, which to a certain extent resists twisting and will only twist so far in response to a given force. From the extent of the twist, one can calculate the amount of difference between the inertial mass and the gravitational mass.

The fiber used was a very thin one so its resistance to twist was very low and yet the horizontal bar showed no measurable turn. Eötvös could calculate that a difference in the two masses of 1 part in 200,000,000 would have produced a measurable twist, so the two masses were identical in amount to within that extent.

(Since then, still more delicate versions of the Eötvös experiment have been carried through and we are now certain by direct observation that inertial mass and gravitational mass are identical in quantity to within 1 part in 1,000,000,000,000.)

Einstein, in working out general relativity, began by assuming that inertial mass and gravitational mass are *exactly* equal because they are, in essence, *the same thing*. This is called "the principal of equivalence" and it plays the same role in general relativity that the constancy of the speed of light plays in special relativity.

It was possible even before Einstein to see that inertially produced acceleration can bring about the same effects as gravitation. Any of us can experience it.

If, for instance, you are in an elevator which starts downward, gaining speed at the start, then during that period of acceleration the floor of the elevator drops out from under you, so to speak, so that you press upon it with less force. You feel your weight decrease as though you were lifting upward. The downward acceleration is equivalent to a lessening of the gravitational pull.

Of course, once the elevator reaches a particular speed and maintains it, there is no longer any acceleration and you feel your normal weight. If the elevator is moving at a constant speed of any amount and in any constant direction, you feel no gravitational effect whatever. In fact, if you are traveling through a vacuum in a totally enclosed box so that you don't see scenery moving, or feel the vibration of air resistance, or hear the whistling of wind, there is absolutely no way you can tell such constant motion from any other (at a different speed or in a different direction) or from being at rest. That is one of the basics of special relativity.

It is because Earth travels through a vacuum at a nearly constant speed and in a nearly constant direction (over short distances) that it is so difficult for people to differentiate the situation from that of Earth being at rest.

On the other hand, if the elevator kept on accelerating downward and moving faster and faster, you would feel as if your weight had decreased permanently. If the elevator accelerated downward at a great enough rate—if it fell at the natural acceleration that gravitational pull would impose upon it ("free fall")—then all sensation of weight would vanish. You would float.

If the elevator accelerated downward at a rate faster than that associated with free fall, you would feel the equivalent of a gravitational pull *upward*, and you would find the ceiling fulfilling the functions of a floor for you.

Naturally you can't expect an elevator to accelerate downward for very long. For one thing, you would need an extraordinarily long shaft within which it might continue moving downward, one that would be light-years long if you want to carry matters to extremes. Then, too, even if you had such an impossibly long shaft, a constant rate of acceleration would soon cause the speed to become a respectable fraction of that of light. That would introduce appreciable relativistic effects and complicate matters.

We can, however, imagine another situation. If an object is in orbit about the Earth it is, in effect, constantly falling toward the Earth at an acceleration imposed upon it by Earth's gravitational pull. However, it is also moving horizontally relative to the Earth's surface, and since the Earth is spherical, that surface curves away from the falling object. Hence the object is always falling, but never reaches the surface. It keeps on falling and falling for billions of years, perhaps. It is in perpetual free fall.

Thus a spaceship that is in coasting orbit about the Earth is held in that orbit by Earth's gravitational pull, but anything on the spaceship falls *with* the spaceship and experiences zero gravity, just as though it were on a perpetually falling elevator. (Actually, astronauts would feel the gravitational pull of the spaceship itself and of each other, to say nothing of the pulls of other planets, and distant stars, but these would all be tiny forces that would be entirely imperceptible.) That is why people on orbiting spaceships float freely.

Again, the Earth is in the grip of the Sun's gravitational pull and that keeps it in orbit about the Sun. So is the Moon. The Earth and the Moon perpetually fall toward the Sun together and, being in free fall, don't feel the Sun's pull as far as their relationship to each other is concerned.

However, the Earth has a gravitational pull of its own which, while much weaker than the Sun's, is nevertheless quite strong. Therefore the Moon, in response to Earth's gravitational pull, moves about the Earth just as though the Sun didn't exist. (Actually, since the Moon is a little removed from the Earth and is sometimes a little closer to the Sun than Earth is, and sometimes a little farther, the Sun's pull is slightly different on the two worlds, and this introduces certain minor "tidal effects" which make evident the reality of the Sun's existence.)

Again, we stand on the Earth and feel only the Earth's pull and not the Sun's at all, since we and the Earth share the free fall with respect to the Sun, and since the tidal effect of the Sun upon ourselves is far too small for us to detect or be conscious of.

Suppose, next, that we are on an elevator accelerating upward. This happens to a minute extent every time we are on an elevator that moves upward from rest. If it is a speedy elevator, then when it starts there is a moment of appreciable acceleration during which the floor moves up toward us and we feel ourselves pressed downward. The acceleration upward produces the sensation of an increased gravitational pull.

Again, the sensation lasts only briefly, for the elevator reaches its maximum speed and then stays there during the course of its trip until it is time for it to stop, when it goes through a momentary slowing and you feel the sensation of a decreased gravitational pull. While the elevator was at maximum speed, and neither speeding up nor slowing down, you would feel perfectly normal.

Well, then, suppose you were in an elevator shaft light-years long and there was an enclosed elevator that could accelerate smoothly upward through a vacuum for an indefinite period, going faster and faster and faster. You would feel an increased gravitational pull indefinitely. (Astronauts feel this for a period of time when a rocket accelerates upward and they are pressed downward uncomfortably. Indeed, there is a limit to how intense an acceleration can be allowed or the additional sensation of gravitational pull can become great enough to press astronauts to death.)

But suppose there is no Earth—just an elevator accelerating upward. If the rate of acceleration were at an appropriate level, you would feel the equivalent of a gravitational pull just like that on Earth's surface. You would walk about perfectly comfortably and could imagine the elevator to be resting motionless on Earth's surface.

Here is where Einstein made his great leap of imagination. By supposing that inertial mass and gravitational mass were identical, he also supposed that there was no way—*no way*—in which you could tell whether you were in an enclosed cubicle moving upward at a steady acceleration of 9.8 meters per second per second, or were in that same enclosed cubicle at rest on the surface of the Earth.

This means that anything that would happen in the accelerating cubicle must also happen at rest on the surface of the Earth.

This is easy to see as far as the falling of ordinary bodies is concerned. An object held out at arm's length in an accelerating cubicle would drop when released and seem to fall at a constantly accelerating rate because the floor of the cubicle would be moving up to meet it at a constantly accelerating rate.

Therefore an object held out on Earth would also fall in the same way. This doesn't mean that the Earth is accelerating upward toward the object. It just means that gravitational pull produces an effect indistinguishable from that of upward acceleration.

Einstein, however, insisted that this included *everything*. If a beam of light were sent horizontally across the upward-accelerating elevator, the elevator would be a little higher when the beam of light flnished its journey and therefore the beam of light would seem to curve downward as it crossed the cubicle. Light travels so rapidly, to be sure, that in the time it took for it to cross the cubicle it would have moved downward only imperceptibly—but it would curve just the same; there is no question about that.

Therefore, said Einstein, a beam of light subjected to Earth's gravitational field (or *any* gravitational field) must also travel a curved path. The more intense the gravitational field and the longer the path traveled by the light beam, the more noticeable the curve. This is an example of a deduction that can be drawn from the principle of equivalence that could not be drawn from earlier theories of the structure of the universe. All the deductions put together make up general relativity.

Other deductions include the suggestion that light should take a bit longer to travel from A to B when subjected to a gravitational field, because it follows a curved path; that light loses energy when moving against the pull of a gravitational field and therefore shows a red shift, and so on.

Again, by considering all the deductions, it makes sense to consider space-time to be curved. Everything follows the curve so that gravitational effects are due to the geometry of space-time rather than to a "pull."

It is possible to work up a simple analogy to gravitational effects by imagining an indefinitely large sheet made up of infinitely stretchable rubber extending high above the surface of the Earth. The weight of any mass resting on that sheet pushes down the rubber at that point and creates a "gravity well." The greater the mass and the more compressed it is, the deeper the wall and the steeper the sides. An object rolling across the sheet may skim one edge of the gravity well, sinking down the shallow rim of the well and out again. In this way it will be forced to follow a curved path just as though it had suffered a gravitational pull.

If the rolling object should happen to follow a path that would take it deeper into the well, it might be trapped and made to follow a slanting elliptical path about the walls of the well. If there is friction between the moving object and the walls, the orbit will decay and the object will eventually fall into the greater object at the bottom of the well.

All in all, making use of general relativity, Einstein was able to set up certain "field equations" that applied to the universe as a whole. Those field equations founded the science of cosmology (the study of the properties of the universe as a whole.)

Einstein announced general relativity in 1916, and the next question was whether it could be verified by observation as special relativity had been soon after its announcement eleven years earlier.

There is a catch. While both special and general relativity predicted effects that differed from the older Newtonian view by so little as to be all but indetectable, the fortuitous discovery of subatomic phenomena made it possible to study very pronounced versions of special relativistic effects.

General relativity had no such luck. For half a century after Einstein had suggested it, there were only very tiny effects that could be relied on to distinguish general relativity from the earlier Newtonian treatment.

Such observations as could be made tended to be favorable to general relativity, but were not overwhelmingly favorable. General relativity therefore remained a matter of dispute for a long time (but *not* special relativity, which is a settled matter).

What's more, because Einstein's version was not strongly borne out, other scientists tried to work out alternate mathematical formulations based upon the principle of equivalence, so that there were a number of different general relativities.

Of all the different general relativities, Einstein's happened to be the simplest and the one that could be expressed most neatly in mathematical equations. It was the most "elegant."

Elegance is powerfully attractive to mathematicians and scientists but it is no absolute guarantee of truth. It was therefore necessary to find tests (if possible) that would not only distinguish Einstein's general relativity from Newton's view of the universe, but from all the competing general relativities.

We'll take that up in the next chapter.

4

THE VICTORIOUS
GENERAL

Carol Brener, the witty proprietress of Murder Ink, a book-
store featuring mystery novels, phoned the other day to ask if
she might send someone over with a copy of my book *The
Robots of Dawn,* so that I might autograph it for a favored
customer. I agreed readily, of course.

The "someone" arrived and, rather to my astonishment,
turned out to be a young lady of considerable beauty. Instantly I was all suavity (as is my wont). I invited her in, and
signed the book.

"Don't tell me," I said to her, fairly oozing charm, "that
Carol sent you into my lair without warning you about me?"

"Oh, she warned me," said the young lady with compo-
sure. "She told me to relax because you were essentially
harmless."

—And it is that, I hope, that is the proper attitude to take
toward this second essay I am writing on general relativity.
The subject may seem formidable but (fingers crossed) let's
hope it will prove essentially harmless.

PHYSICS

* * *

In the preceding chapter I explained that general relativity was based on the assumption that gravitational mass was identical with inertial mass, and that one could therefore look upon gravitational effects as identical to the effects one would observe in an endlessly accelerating system.

The question is: How can one demonstrate that this view of gravitation is more correct than Newton's is?

To begin with, there are what have been called "the three classic tests."

The first of these arose out of the fact that at the time general relativity was advanced by Einstein in 1916, there remained one nagging gravitational puzzle concerning the solar system. Every time Mercury revolved about the Sun in its elliptical orbit, it passed through that point at which it was closest to the Sun ("perihelion"). The position of this perihelion was not fixed relative to the background of the stars, but advanced slightly at each turn. It was expected to do so because of the minor effects ("perturbations") of the gravitational pulls of other planets. However, when all these perturbations were taken into account, there was a slight advance of the perihelion left over, one that amounted to forty-three arc seconds per century.

That was a very small motion (it would amount to only the apparent width of our Moon after 4,337 years) but it was detectable and bothersome. The best explanation that could be advanced was that an undiscovered planet existed inside Mercury's orbit and that its unallowed-for gravitational pull supplied the reason for the otherwise unexplained advance of the perihelion. The only catch was that no such planet could be detected (see "The Planet That Wasn't" in *The Planet That Wasn't*, Doubleday, 1976).

To Einstein, however, the gravitational field was a form of energy, and that energy was equivalent to a tiny mass, which in turn produced an additional bit of gravitational field. Therefore the Sun had a little bit more gravitation than was credited to it by Newtonian mathematics and *that*, not another planet, just accounted for the advance of Mercury's perihelion.

This was an instant and impressive victory for general relativity, and yet the victory proved to have its limitations. All the calculations that dealt with the position of Mercury's perihelion included the assumption that the Sun was a perfect sphere. Since the Sun is a ball of gas with a very intense gravitational field, this seems a fair assumption.

However, the Sun did rotate and, as a result, it should be an oblate spheroid. Even a small equatorial bulge on the Sun might produce an effect that would account for part or all of the advance, and this would cast doubt on general relativity.

In 1967 the American physicist Robert Henry Dicke made delicate measurements of the size of the solar disc and reported a slight oblateness that was sufficient to account for three of the forty-three arc seconds of advance per century. It made scientific headlines as a possible blow to Einstein's general relativity.

Since then, however, there have been reports of smaller values of solar oblateness and the matter remains in dispute. My own feeling is that in the end it will turn out that the Sun is only insignificantly oblate, but right now the advance of Mercury's perihelion is not considered a good test of Einstein's general relativity.

What about the other two classic tests?

One of them involved the matter of light curving in a gravitational field, something I mentioned in Chapter 3. If this actually took place to the amount predicted by general relativity, it would be more impressive than the matter of Mercury's perihelion. After all, the motion of Mercury's perihelion was known, and Einstein's mathematics might conceivably have been worked out to fit it. On the other hand, no one had ever thought to test for gravitational curving of light because, for one thing, no one had dreamed such a phenomenon might exist. If such an unlikely phenomenon were predicted and if it then turned out to exist, that would be an incredible triumph for the theory.

How to test it? If a star happened to be located very near the position of the Sun in the sky, its light, skimming past the Sun, would curve in such a way that the star would seem to be located slightly farther from the Sun's position than it really was. General relativity showed that a star whose light just grazed the solar edge would be displaced by 1.75 arc seconds, or just about a thousandth of the apparent width of the Sun. This is not much, but it is measurable—except that those stars that are close to the Sun's apparent position in the sky aren't ordinarily visible.

During a total eclipse of the Sun they would be, however, and such an eclipse was scheduled for May 29, 1919. As it happened, the darkened Sun would then be located amid a group of bright stars. The British astronomer Arthur Stanley Eddington, who had managed to get a copy of Einstein's paper on general relativity through the neutral Netherlands during the dark days of World War I, was impressed by it and organized an expedition to make the necessary measurements of the positions of those stars relative to one another. Such measurements could then be compared with the known positions of those same stars at times when the Sun was far distant in the sky.

The measurements were made and, to the growing excitement of the astronomers, star after star showed the predicted displacement. General relativity was demonstrated in a manner that was incredibly dramatic, and the result was front-page news. At a bound, Einstein became what he remained for the rest of his life—the most famous scientist in the world.

And yet, although the 1919 eclipse is supposed (in the popular mythology of science) to have settled the matter, and although I, too, have always treated it as having done so, in actual fact it did not really establish general relativity.

The measurements were necessarily fuzzy, the comparison between those measurements and the positions at other times of the year were hard to narrow down with precision, additional uncertainty was introduced owing to the fact that at different times of the year different telescopes were used under different weather conditions, and, on the whole, as support for general relativity the data was shaky. It certainly could not serve to distinguish Einstein's variety from the other, competing varieties that were eventually offered.

What's more, later measurements in later eclipses did not seem to improve the situation.

And the third of the classic tests?

I mentioned in Chapter 3 that light rising against the pull of gravitation should lose energy, according to general relativity, since light would certainly do so if it rose against an upward acceleration of the source. The loss of energy meant that any spectral line that was at a given wavelength in the absence of a significant gravitational field would shift toward the red if the light containing it moved against the pull of gravity. This was the "gravitational redshift," or the "Einstein redshift."

Such a redshift, however, was very small indeed, and it would take an enormously intense gravitational field to produce one that could be unmistakably measured.

At the time Einstein advanced his theory of general relativity, the most intense gravitational field that could be easily studied seemed to be that of the Sun, and that, intense as it was, was too weak to be useful as a test for the Einstein redshift.

However, just a matter of months before Einstein's paper,

the American astronomer Walter Sydney Adams had produced evidence that Sirius's dim companion ("Sirius B") was actually a star with the mass of the Sun but the volume of only a small planet (see "How Little?" in *The Sun Shines Bright*, Doubleday, 1981). This was a little hard to believe at first, and for a while the concept of the "white dwarf" was ignored.

It was Eddington, however, who saw, quite clearly, that if Sirius B was indeed very small it had to be very dense, and should have an enormously intense gravitational field. Its light should therefore show a clearly perceptible Einstein redshift if general relativity were correct.

Adams went on to study the spectrum of Sirius B in detail, and in 1925 reported that the Einstein redshift was there and was rather close to what general relativity predicted.

Again this was hailed as a triumph, but again, once the period of euphoria passed, it seemed there was a certain fuzziness to the result. The measurement of the shift was not very accurate for a number of reasons (for instance, the movement of Sirius B through space introduced a spectral-line shift that was unrelated to general relativity and that introduced an annoying uncertainty). As a result, the test certainly couldn't be used to distinguish Einstein's general relativity from other competing theories, and the study of light issuing from other white dwarfs didn't seem to improve matters.

As late as 1960, then, forty-four years after general relativity had been introduced and five years after Einstein's death, the theory still rested on the three classic tests that were simply inadequate for the job. What's more, it seemed as though there might not be any other test that could even begin to settle the matter.

It looked as though astronomers would simply have to live without an adequate description of the universe as a whole and argue over the various general relativistic possibilities forever, like the Scholastics debating the number of angels who could dance on the head of a pin.

About the only thing one could say, constructively, was that Einstein's version was the simplest to express mathematically and was, therefore, the most elegant—but that was no certain argument for truth either.

Then, from 1960 onward, everything changed—

The German physicist Rudolf Ludwig Mössbauer received his Ph.D. in 1958 at the age of twenty-nine, and in that same year announced what has come to be called the "Mössbauer effect," for which he received the Nobel Prize for physics in 1961.

The Mössbauer effect involves the emission of gamma rays by certain radioactive atoms. Gamma rays consist of energetic photons, and their emission induces a recoil in the atom that does the emitting. The recoil lowers the energy of the gamma-ray photon a bit. Ordinarily the amount of recoil varies from atom to atom for a variety of reasons, and the result is that when the photons are emitted in quantity by a collection of atoms, they are apt to have a wide spread in energy content.

There are, however, conditions under which atoms, when they exist in a sizable and orderly crystal, will emit gamma-ray photons with the entire crystal undergoing the recoil as a unit. Since the crystal is enormously massive as compared with a single atom, the recoil it undergoes is insignificantly small. All the photons are emitted at full energy, so the beam has an energy spread of virtually zero. This is the Mössbauer effect.

Gamma-ray photons of *exactly* the energy content emitted by a crystal under these conditions will be strongly absorbed by another crystal of the same type. If the energy content is even very slightly different in either direction, absorption by a similar crystal is sharply reduced.

Well, then, suppose a crystal that is emitting gamma-ray photons is in the basement of a building and the stream of photons is allowed to shoot upward to an absorbing crystal on the roof, 20 meters above. According to general relativity, the photons climbing against the pull of Earth's gravity would lose energy. The amount of energy they lost would be extremely small, but it would be large enough to prevent the crystal on the roof from absorbing it.

On March 6, 1960, two American physicists, Robert Vivian Pound and Glen Rebka, Jr., reported that they had conducted just this experiment and found that the photons were *not* absorbed. What's more, they then moved the receiving crystal downward very slowly so that its motion would very slightly increase the energy of collision with the incoming photons. They measured the rate of downward movement that would cause just enough of an energy increase to make up for the general relativistic loss and to allow the photons to be strongly absorbed. In this way they determined exactly how much energy was lost by the gamma rays in climbing against Earth's gravitational pull, and found that the result agreed with Einstein's prediction to within 1 percent.

This was the first real and indisputable demonstration that general relativity was correct, and it was the first demonstration that was held entirely in the laboratory. Until then, the three classic tests had all been astronomical and had required measurements with built-in uncertainties that were almost impossible to reduce. In the laboratory, everything could be tightly controlled, and precisions were much higher. Astonishingly, too, the Mössbauer effect did not require a white dwarf, or even the Sun. The Earth's comparatively feeble gravitational field sufficed, and over a difference of height amounting to no more than that between the basement and roof of a six-story building.

However, although the Mössbauer effect might be considered as having nailed general relativity into place at last, and as having definitely outmoded Newtonian gravity, the other varieties of general relativity (that were introduced after 1960, in fact) were not eliminated by this experiment.

On September 14, 1959, a radar echo was received for the first time from an object outside the Earth-Moon system—from the planet Venus.

Radar echoes are produced by a beam of microwaves (very high-frequency radio waves) that travel at the speed of light, a figure we know with considerable precision. A microwave beam will speed to Venus, strike its surface and be reflected, and will return to Earth in anywhere from 2¼ to 25 minutes depending on where Earth and Venus are in their respective orbits. From the actual time taken by the echo to return, the distance of Venus at any given time can be determined with a precision greater than any earlier method had made possible. The orbit of Venus could therefore be plotted with great exactness.

That reversed the situation. It became possible to predict just how long it should take a beam of microwaves to strike Venus and return when the planet was at any particular position in its orbit relative to ourselves. Even slight differences from the predicted length of time could be determined without serious uncertainty.

The importance of this is that Venus at 584-day intervals will be almost exactly on the opposite side of the Sun from ourselves, so the light streaking from Venus to Earth must skim the Sun's edge on the way.

By general relativity, that light should follow a curved path and the apparent position of Venus should shift very slightly away from the Sun. But Venus can't be observed when it is that close to the Sun, and even if it could be, the slight shift in position would be next to impossible to measure with confidence.

However, because light follows a slightly curved path in skimming the Sun's surface, *it takes longer to reach us* than if it had followed the usual straight line. We can't measure the time it takes Venus-light to reach us, but we can send out a microwave beam to Venus and wait for the echo. The beam will pass near the Sun as it moves each way and we can measure the time delay in receiving the echo.

If we know how closely the microwave beam approaches the Sun, we should know, by the mathematics of general relativity, exactly by how much it should be delayed. The actual delay and the theoretical can be compared with far more certainty than we can measure the displacement of the stars at the time of a total eclipse.

Then, too, our planetary probes emit microwave pulses and these can be detected. As the precise distance of the probe at any moment is known, the time taken for the pulses to travel to the Earth can be measured and compared to the theoretical, when the pulses move nowhere near the Sun, and then again when they must skim past the Sun. Such measurements, made from 1968 onward, have shown agreement with Einstein's formulation of general relativity to within 0.1 percent.

There seems to be little doubt now, therefore, that not only is general relativity correct, but that it is Einstein's formulation that is the victorious general. Competing theories are fading out of the field.

There are now astronomical demonstrations, too, of the validity of general relativity, demonstrations that involve objects that were not known to exist at the time Einstein first advanced his theory.

THE VICTORIOUS GENERAL

In 1963 the Dutch-American astronomer Maarten Schmidt managed to demonstrate that certain "stars" that were strong emitters of radio waves were not stars of our own galaxy, but were objects located a billion or more light-years away. This could be demonstrated by the enormous redshift of their spectral lines, which showed they were receding from us at unprecedentedly high speeds. This could only be so (presumably) because they were so vastly far away from us.

This aroused considerable controversy over what these objects ("quasars") could possibly be, but that controversy is of no importance in connection with our present concern. What is important is that quasars emit strong beams of radio waves. Thanks to the elaborate radio telescopes built since quasars were first recognized for what they are, the radio sources within quasars can be located with pinpoint accuracy far greater than that with which it is possible to locate a merely light-emitting object.

Occasionally the light (and radio) waves from a particular quasar skim the Sun's surface on their way to us. The light waves are lost in the Sun's overwhelming brilliance but the radio waves can be easily detected, Sun or not, so there is no need to wait for an eclipse that just happens to take place when the Sun is in the right position for our purposes. What's more, the radio-wave source is so accurately recorded that the slight shift induced by general relativity can be much more accurately determined than the famous shift in star position during the 1919 eclipse.

The shift of position in quasar radio waves, measured a number of times over the last fifteen years, proves to be within less than 1 percent of what Einstein's general relativity predicts it should be, and the measurements made during the 1919 eclipse, shaky and uncertain as they were, have been vindicated.

Quasars are involved in another phenomenon that supports general relativity, one that is particularly dramatic.

Suppose there is a distant light-emitting object and between it and ourselves is a small object with a powerful gravitational field. The distant light-emitting object would send out light waves that would skim past the unseen nearer object on all sides. On all sides the light would be displaced outward by the general relativity effect, and the result would be exactly as though light were passing through an ordinary glass lens. The distant object would be magnified and made to seem larger than it really was. This would be a "gravitational lens" and its existence was early predicted by Einstein himself.

The trouble with the concept was that no known case of it existed in the sky. There was no large luminous star, for instance, that had a tiny white dwarf right between itself and us. Even if there were, how could we tell that the star was slightly expanded from what it would normally be if the white dwarf weren't there? We couldn't remove the white dwarf and watch the star shrink back to its normal size.

But consider quasars. The quasars are all farther away than ordinary galaxies, and ordinary galaxies number in the billions. There is a reasonable possibility that there might be a small galaxy lying between ourselves and one of the hundreds of quasars now known. What's more, the radio source within a quasar (which is what we most accurately observe) and the intermediate galaxy would both be irregular objects, so the effect would be similar to that of light passing through a badly flawed lens. Instead of just expanding, the quasar might break up into two or more separate images.

In 1979 a team of American astronomers, D. Walsh, R. F. Carswell, and R. J. Weymann, were observing quasar 0957 + 561, which showed two radio sources about 6 arc seconds apart. They seemed to be two quasars that were equally bright and equally distant from us. What's more, their spectra seemed to be identical. The astronomers suggested that what was being observed was really a single quasar that was split in two through a gravitational-lens effect.

The vicinity of the quasar was examined very closely for any sign of galaxies between ourselves and it, and in 1980 it was shown that there was a cluster of faint galaxies at about a third of the distance of the quasars and just in front of them. The conditions seemed to be just right for the production of a gravitational lens, and since then other possible cases have been discovered—another score for general relativity.

But the most dramatic and important demonstration of general relativity remains to be told.

Einstein predicted the existence of gravitational waves analogous to light waves. Accelerating masses should emit gravity waves, just as oscillating electromagnetic fields emit light waves and similar radiation. Thus any planet circling our Sun is continually changing direction as it circles, and is therefore continually accelerating. It should be emitting gravitational waves, losing energy in consequence, approaching the Sun and eventually falling in. This is happening to Earth, for instance, but the loss of energy is so small that it is hopeless to expect to detect the effect.

What is needed are more intense gravitational fields and more extreme accelerations. Nothing even approaching what was necessary was known until 1974.

In that year the American astronomers Russell A. Hulse and Joseph H. Taylor, Jr., discovered a pulsar that is now called PSR1913 + 16. It emitted radio-wave pulses at intervals of 0.05902999527 seconds, or just about 17 pulses per minute. These intervals grow slightly larger and slightly smaller in a regular way with a period of 7.752 hours.

The implication is that it is alternately approaching us and receding from us, and the best way of explaining that is to suppose that it is revolving about something. From the size of its orbit and the fact that the object it circles can't be seen, the astronomers concluded that they had a double pulsar by the tail.

This in itself is not unprecedented. Other double pulsars have been located. What is unusual, however, is that the two pulsars of this system are so close together. They whizz around each other at speeds of nearly 200 miles per second. That, combined with the smallness of the orbit and the intensity of their gravitational fields, meant that general relativistic effects should be enormous.

For example, the point of closest approach of the pulsars to each other ("periastron") should move forward just as Mercury's perihelion does, but at better than one and a half million times the rate. And sure enough, the advance has been noted at an appropriate rate of 4.226 degrees per year.

Even more important, the binary pulsar should give off floods of gravitational waves in quantity sufficient to shorten the period of revolution perceptibly.

The shortening should be only a ten millionth of a second per orbital period. This accumulates, however, as the number of orbits through which it is observed mounts up, and by now there is no doubt that the pulsars of the system are shortening their orbits and approaching each other, and that in less than ten thousand years they should crash into each other.

This is strong evidence in favor of the gravitational waves predicted by Einstein's general relativity theory.

And that's the story. Every appropriate measurement that has been made in two thirds of a century has supported Einstein. Not one measurement has successfully cast any serious doubt upon him.

I would be sorry that Einstein hadn't lived long enough to see at least some of the victories that have taken place since 1960, but it really doesn't matter. He was always supremely sure of the correctness of his theory. There is a story that after the 1919 eclipse he was asked what he would have thought if the star-displacement measurements had not supported him. He is supposed to have answered that he would have been sorry for God for having made the mistake of building the universe on the wrong principles.

PART II
ASTRONOMY

5

UPDATING THE
SATELLITES

As one gets older, one tends to collect a variety of reputations. About the best one I've got is that of being a "nice person."

That's a pleasant reputation to have because it means that people grin at me, and pump my hand, and massage my shoulder, and let their eyes glisten with pleasure on meeting me. Young women of surpassing beauty are even likely to ask permission to kiss me.*

Nevertheless, it is a little wearying sometimes to have editors protect their property by not letting me snarl and swear and grind my teeth when I have an urgent need to.

Consider my book *Saturn and Beyond* (Lothrop, Lee and Shepard, 1979). At the time I wrote it, Pluto was not known to possess a satellite. By the time the galleys arrived, the satellite had been discovered and I dutifully added a couple of paragraphs so that when the book appeared, the Plutonian satellite was properly ensconced within its pages.

---

*I never withhold such permission. That would be rude!

Some time after publication there arrived a review of the book that castigated me unmercifully for not including the satellite. The tone of the review was insulting in the extreme.

My editor did a little detective work and it turned out that the reviewer had read galley proofs plainly marked "uncorrected," and was too stupid to understand that one of the purposes of galley proofs is to give the author a chance to bring material up to date.

The editor didn't want me to write to the reviewer directly for some reason. (Perhaps she knew something about the nature of my eloquence.) She suggested instead that she should be allowed to intercept the letter, and then forward it to the reviewer.

I agreed, and promptly knocked out a letter explaining the situation in detail. I began with a short, Ciceronian essay on the subject of "stupidity," and then went on to discuss the symptoms and consequences of "senility," and concluded with some pleasant suggestions as to what the reviewer could do with various parts of his body.

Unfortunately (I know you will scarcely credit this), my editor refused to pass it on, but sent a milk-and-water letter of her own which left the reviewer totally unscathed.

Her excuse was that my letter didn't project my "nice person" image correctly. My heated explanation that I didn't feel like a nice person at all, but wanted to eviscerate the son of an uncertain father, fell on deaf ears.

But never mind—my books do get outdated with time, and one of the ways in which *Saturn and Beyond* got outdated (as well as an earlier sister volume, *Jupiter, the Largest Planet* (Lothrop, 1973) was in connection with the satellites of the solar system. Here's my chance to update such matters.

If we begin at the Sun and work outward, it turns out that Mercury and Venus have no satellites at all, as far as we know, and it seems quite certain that none will ever be discovered of any significant size at all.

Earth has one satellite, the Moon, and it seems quite certain that no second satellite (of natural origin) exists that is of any significant size. The Moon is a large satellite, however— one of a total of seven in the solar system that have diameters in excess of 3,000 kilometers. It is very unlikely (barring the discovery of a gas giant beyond Pluto's orbit) that any large satellites remain to be discovered. In the last century and a third, only one new satellite (which I will get to) has been spotted that is over 200 kilometers in diameter.

Mars has two tiny satellites that have been known for a century, and that recently have been photographed in detail.

That brings us to Jupiter. It has four large satellites, sometimes called the "Galilean satellites" because Galileo discovered them in 1610. One of these, Ganymede, is the largest satellite in the solar system, with a volume 3½ times that of the Moon.

In addition to the Galilean satellites Jupiter has a number of small satellites, and here is where the updating begins.

In 1973 when my book *Jupiter* was published, eight small Jovian satellites were known. (I have the impulse to call them "satellettes" but I am resisting it.) The Jovian satellites are frequently numbered in the order of discovery, with the Galileans J-I to J-IV, and the small ones J-V to J-XII (as of 1973). The "J" stands for Jupiter as you have already guessed.

In *Jupiter*, I gave the eight small satellites names drawn from mythology. These names were unofficial at the time, but I assumed they would become official. I was wrong. Only one of the eight was kept, so I must start my updating by giving you the now official names of the eight small satellites, along with the year of discovery:

J-V	Amalthea (am-ul-THEE-uh)	1892
J-VI	Himalia (HIM-uh-lee-uh)	1904
J-VII	Elara (EE-luh-ruh)	1905
J-VIII	Pasiphaë (puh-SIF-uh-ee)	1908
J-IX	Sinope (sy-NO-pee)	1914
J-X	Lysithea (LIS-ih-THEE-uh)	1938
J-XI	Carme (KAHR-mee)	1938
J-XII	Ananke (UH-nan-kee)	1951

Traditionally the names of bodies of the solar system are taken from Greek mythology, and these are no exceptions.

Amalthea (the only name I used earlier that was kept and finally made official) was one of the nymphs who fed the infant Zeus (Jupiter, to the Romans) on goat's milk when he was hidden in Crete to keep him safe from the cannibalistic tendencies of his father, Cronos (Saturn). Sometimes the name is given to the goat who supplied the milk. In either case it is appropriate for the satellite that was closer than any other known Jovian satellite at the time it was discovered.

When Zeus matured, by the way, he gave the horn of the goat to the nymph as a reward, telling her that if she wanted anything she had but to reach into the horn to find it. (This was the original cornucopia, from a Latin phrase meaning "horn of plenty.")

Elara was a mortal woman who caught the eye of the all-embracing Zeus. He hid her underground to keep her from being discovered by the jealous Hera (Juno) who never did get used to the amorous propensities of her all-powerful husband, and who got her revenge by persecuting anyone he seemed to like.

According to some accounts Elara was the mother of Tityus, a huge Earth-born monster (Elara was underground, remember) who was killed by the arrows of Apollo and who, when stretched out in Tartarus, covered nine acres of ground.

Pasiphaë was a daughter-in-law of Zeus, having married his son, King Minos of Crete. Pasiphaë is best known for having fallen violently in love with a bull of great beauty. (There's no accounting for tastes.) She had a framework made, and over it a cow's hide was stretched. Pasiphaë got inside and soon enough the bull obligingly mounted the framework. In due course Pasiphaë was delivered of a child with a bull's head that grew into the famous Minotaur.

Sinope was another young lady who was approached by the insatiable Zeus. He offered her anything she wanted in exchange for her compliance and she asked for perpetual virginity. (I told you there's no accounting for tastes.)

Carme was still another Zeusian beneficiary, and the mother of Britomartis, a Cretan patron-goddess of fishing and hunting.

Ananke differs from the rest. She is the divine personification of Fate or Necessity—the ordained playing-out of events which even the gods cannot alter, so that Ananke is the one divinity that is superior to Zeus.

Himalia and Lysithea are quite obscure figures, which I tracked down only through the kindness of several of my readers. (Astronomers are either unusually knowledgeable about mythology, or are desperate enough to hunt through large compendia.)

In any case, Himalia was a nymph of Rhodes, who presided over rich harvests and who bore good old Zeus another rich harvest, three sons.

Lysithea is a nymph who in some myths is identified as the mother of Dionysos, god of the vine. The mother he is usually given is Semele.

The eight small Jovian satellites were discovered in the order of decreasing brightness, as seems reasonable. Since all are at the same distance from Earth and are probably of similar albedo (the ability to reflect light), they were discovered in the order of decreasing size. Thus, Amalthea and

Himalia have diameters of about 170 kilometers, while Ananke has a diameter of perhaps 20 kilometers.

But even the largest is comparatively tiny. All eight satellites make up only $1/3,000$ the volume of Europa, the smallest of the Galileans, or $1/32,000$ the volume of the four Galilean satellites put together.

Four of the small satellites are grouped together at a particularly great distance from Jupiter. They are Pasiphaë, Sinope, Carme, and Ananke, with average distances from Jupiter ranging from 20,700,000 kilometers in the case of Ananke, the nearest, to 23,370,000 kilometers in the case of Sinope, the farthest.

These satellites are very likely captured asteroids and are relatively recent additions to the Jovian family. They have not yet had time to regularize their orbits into circularity and to move into Jupiter's equatorial plane—especially since, at their distance, Jupiter's gravitational influence is comparatively weak. The satellite orbits are therefore strongly tilted and elliptical, and the grouping is by no means as tight as it would be if all had circular orbits and revolved in Jupiter's equatorial plane.

The most eccentric orbit is that of Pasiphaë, which is rather suitable considering the eccentric sexual tastes of the mythological prototype. Pasiphaë recedes to a distance of 33,200,000 kilometers from Jupiter at one end of its orbit. This is the farthest known distance of any satellite from the planet it circles. It is roughly 85 times the distance of the Moon from the Earth, for instance.

The periods of revolution of these satellites are long, well over 600 days in each case. The longest period of revolution is that of Sinope, as would be expected, since it has the greatest average distance from Jupiter. Its period is 758 days, or 2.08 years. This is nearly 28 times as long as it takes the Moon to circle the Earth, and is 1.1 times as long as it takes Mars to circle the Sun.

UPDATING THE SATELLITES

Of the remaining small satellites, three are grouped some-what closer to Jupiter. Himalia, Lysithea, and Elara have average distances of between 11,000,000 and 12,000,000 kilometers from Jupiter.

There is an overlap, though. Elara has an orbit that is eccentric enough for it to recede as far as 14,300,000 kilometers from Jupiter, while Pasiphaë at the near end of its orbit is only 13,800,000 kilometers from Jupiter.

All seven satellites are much farther from Jupiter than the Galilean satellites are. The nearest approach of any of the seven is that of Elara which, at the near point of its orbit, is only 9,300,000 kilometers from Jupiter. This, however, is five times as far from Jupiter as Callisto, the farthest of the Galileans, ever gets.

The eighth small satellite, Amalthea, differs from all the rest in being *closer* to Jupiter than any of the Galileans are. It is at a distance of only 180,000 kilometers from the center of Jupiter. This is not quite half the distance of Io, the closest of the Galileans, and, for that matter, not quite half the distance of the Moon from the Earth.

Impelled by the enormous gravitational field of nearby Jupiter, Amalthea is whipped about the planet in 11.95 hours, which is less than $\frac{1}{50}$ the time it takes for the Moon to circle the Earth.

At the time I wrote *Jupiter*, Amalthea had the second shortest period of any known satellite. Amalthea was beaten only by Phobos, the inner of Mars's two satellites. It revolves about Mars in 7.65 hours, only $\frac{5}{8}$ the period of Amalthea. Phobos revolves about Mars, in fact, considerably faster than Mars rotates about its axis, so that Phobos constantly over-takes the Martian surface, rising in the west and setting in the east. Since Jupiter rotates surprisingly quickly, in 9.92 hours, Amalthea does not overtake the Jovian surface but rises in the east and drifts rather slowly to the west.

Phobos, however, as it circles tiny Mars, has a much shorter orbit than that of Amalthea, which has to get about the swollen globe of mighty Jupiter. Amalthea's orbit is nearly twenty times as long as that of Phobos. Therefore Phobos, in its orbit about Mars, moves at a speed of 2.14 kilometers per second while Amalthea, scudding about Jupiter, moves at a speed of 26.3 kilometers per second. (By comparison, the Moon moves about Earth at an average speed of just about 1 kilometer per second.)

But *Jupiter* was published in 1973, and in 1974 a thirteenth Jovian satellite was discovered by Earth-based observation. It was part of the Himalia group, as the small satellites outside the Galileans are referred to, but not all the way outside. That increased the number of that group from three to four.

The reason this new satellite was not discovered any sooner was that it was the smallest yet. It was, in fact, only 10 kilometers across, and it remains to this day the smallest satellite to have been found.

It was named Leda (LEE-duh), who, in Greek mythology, was a queen of Sparta who was wooed by Zeus. The god took the shape of a swan for the purpose, thus giving occasion for a number of artistic representations of bestiality. The result was that Leda laid two eggs and from each of these, two babies emerged. The most famous of the babies was one who eventually came to be known as Helen of Troy.

Then came the age of the probes, and in 1979 three new Jovian satellites were discovered, all of which were closer to Jupiter than the Galilean satellites were. This made for a rather startling symmetry. There are now 16 Jovian satellites: 4 small ones closer to Jupiter than the Galileans are; 4 large Galileans; 4 small ones farther than the Galileans; and 4 more small ones still farther out. Undoubtedly there will be a new discovery that will break the symmetry, and I would consider that a shame, because I like symmetry.

The most recently discovered satellites, Adrastea, Thebe, and Metis,† have estimated diameters of about 25, 80, and 40 kilometers respectively, and one might wonder why it took so long to discover them when Lysithea, with a diameter of only 20 kilometers, was discovered as long ago as 1938.

The answer is that these nearby satellites are drowned out in the light of giant Jupiter and cannot be seen except by the close-in view of the probes. Amalthea, the only nearby satellite to be detected from Earth, is 170 kilometers across, far larger than the others, and was discovered in 1892 by an astronomer of almost legendary sharpness of vision.

The nearest of all the satellites to Jupiter (as far as is now known) is Metis, which is only 128,000 kilometers from the planet's center, though Adrastea is a close second at 129,000 kilometers. The period of revolution of Metis is 7.07 hours, while that of Adrastea is 7.13 hours. They have taken the record away from Phobos, since they negotiate their trip around their planet in half an hour less than Phobos does.

The orbital speed of these two innermost Jovian satellites is a quick 31.6 kilometers per second, and both outrace the rotation of Jupiter. If anyone could watch those satellites from the Jovian cloud surface they would seem, like Phobos, to rise in the west and set in the east.

---

†Adrastea is the name of a nymph who took care of the infant Zeus, along with Amalthea. Thebe was a nymph who was the daughter of a river-god and, you guessed it, Zeus saw her once and what followed was inevitable. Metis is the most interesting of the three, for she was a Titaness, one of the older generation of deities, who, from her name, was personified wisdom. She is considered to have been Zeus's first wife, but he feared that she might give birth to a son who would depose him, as Zeus had deposed *his* father, Kronos. Zeus therefore swallowed Metis, who thereafter counseled him from his stomach (one way to get an education). Nine months later Athena burst forth from Jupiter's head, fully armed and shouting her war cry.

* * *

Now let's move on to Saturn. Through the first seven decades of the twentieth century, it was thought to have nine satellites. One of them, the sixth counting outward from Saturn, is a large satellite named Titan. (In volume it is second only to Ganymede, and is even more remarkable in that it is the only satellite known to have an atmosphere—and one that is thicker than that of Earth, too.) Titan has about ten times the volume of all the other Saturnian satellites taken together.

The other Saturnian satellites, while considerably smaller than Titan, are all larger than any of the Jovian satellites other than the Galileans. Rhea (REE-uh) has a diameter of 1,530 kilometers, for instance, and that of Iapetus (i-AP-uh-tus) is 1,460 kilometers. The smallest of the nine Saturnian satellites is Phoebe (FEE-bee), which is the farthest out and, not surprisingly considering its size, was the latest discovered. It is about 220 kilometers in diameter and was first spotted in 1898.‡

Why does Saturn lack the really small satellites that Jupiter has in profusion? The obvious explanation is that Saturn is twice as far from us as Jupiter is, and that the smaller satellites would therefore be correspondingly more difficult to see. They exist, probably, but remain undetected.

In 1967 a small Saturnian satellite was reported and named Janus, something I described in "Little Found Satellite" in *The Solar System and Back* (Doubleday, 1970). Alas, it turned out to be an error, and in this essay I bring myself up to date in that matter—cancel Janus!

‡The names of the Saturnian satellites are given and explained in "Rollcall" in *Of Time and Space and Other Things*, (Doubleday, 1965). The names of the satellites of the planets beyond Saturn are also given there—with one exception.

UPDATING THE SATELLITES

In 1980, however, probes photographing Saturn from close up located no fewer than eight new Saturnian satellites, every one of them smaller than Phoebe. The largest of the eight is slightly less than 200 kilometers in diameter, while the smallest are only about 15 kilometers in diameter, on the average. None of the eight has yet been given a name.

Five of the eight newly discovered Saturnian satellites are closer to Saturn than Mimas (MI-mus) is. (Mimas is the closest of the long-established satellites, having been detected first in 1789, and is some 390 kilometers in diameter.)

The closest of the now-known Saturnian satellites is only 137,000 kilometers from Saturn's center (not quite as close as the closest Jovian satellites are to their planet). It circles Saturn in 14.43 hours. This is twice the period of the closest Jovian satellites, but then Saturn, being smaller than Jupiter, has a less intense gravitational field.

Of the five nearby Saturnian satellites, the two least nearby offered something astonishing and, in fact, until then unprecedented. They are co-orbital—that is, they share the same orbit, chasing each other around Saturn endlessly. They are at a distance of 151,000 kilometers from Saturn and revolve in a period of 16.68 hours. It was these two satellites, mistaken for a single body, that were probably reported as Janus in 1967.

The three remaining newly discovered Saturnian satellites, which occur within the system of the nine long-known satellites, represent other unprecedented situations.

The long-known satellite Dione (di-O-nee), first spotted in 1684, was found to have an unsuspected tiny co-orbital companion. Whereas Dione has a diameter of 1,120 kilometers, the companion (which it would make sense to call Dione-B) has a diameter of only about 30 kilometers. Dione-B circles Saturn at a point 60 degrees ahead of Dione, so that Saturn,

Dione, and Dione-B are at the apexes of an equilateral triangle. This is a "Trojan situation" (see "The Trojan Hearse" in *View from a Height*, Doubleday, 1963).

Until 1980, the only known examples of the Trojan situation involved the Sun, Jupiter, and certain asteroids which were co-orbital with Jupiter. Some of these asteroids circled the Sun 60 degrees ahead of Jupiter in the "L-4" position, and some 60 degrees behind Jupiter in the "L-5" position.

Dione-B is in the L-4 position with respect to Dione.

Even more astonishing is the case of Tethys (TEE-this), which was discovered in the same year as Dione and is 1,060 kilometers in diameter. The two remaining newly discovered Saturnian satellites, each 25 kilometers across, are *both* co-orbital with Tethys. One, Tethys-B, is in the L-4 position with respect to it, and the other, Tethys-C, is in the L-5 position.

Clearly, the Saturnian satellite family is the richest and most complex in the solar system as far as we now know. This is probably part of the same phenomenon that gives Saturn the most spectacular rings in the solar system.

Uranus and Neptune have had no new satellite discoveries in the last third of a century (not having been probed yet) though Uranus's rings have recently been discovered (see "Rings and Things" in *The Road to Infinity* (Doubleday, 1979).

Uranus has five satellites. Four have been known for over a century, and have diameters in the 1,000 to 2,000 kilometer range. The fifth is Miranda (mih-RAN-duh), closer and smaller than the others. It was first detected in 1948, has a diameter of about 300 kilometers, and circles Uranus at a distance of about 130,000 kilometers.

Neptune has two satellites. One is a large satellite, Triton (TRI-tun), with a diameter of about 4,400 kilometers, so that it is larger than our Moon. It was detected only a few weeks after Neptune itself was discovered.

Neptune's second satellite, Nereid (NEER-ee-id), was discovered in 1949 and also has a diameter of about 300 kilometers. Nereid is remarkable for having the most eccentric orbit of any satellite. At one end of its orbit it swoops in as close as 1,390,000 kilometers to Neptune, and at the other end recedes to a distance of 9,734,000 kilometers.

What's more, the latest figure I can find on the orbital period of Nereid makes that out to be 365.21 days, or 0.9999 of a year.

Imagine what might have happened if Neptune and Nereid were visible to the unaided eye and human beings could watch the latter circling the former. It wouldn't have taken long for even prehistoric men to realize that Nereid was marking off the cycle of the seasons with great precision.

We would have ended up with a nice Neptunian calendar, complete with leap years, long before the Neolithic. Heaven only knows how that would have stimulated mathematics, science, and technology, and where we would now be as a result.

If that had happened, the existence of Nereid would have been considered a clear example of the benign providence of God, and scientists would have been hard put to it to say "Oh, it's just a coincidence." However, since benign providence arranged to have this astonishing coincidence invisible until our present generation, the matter does not arise.

Oddly enough, the most astonishing satellite discovery of the last few years has had nothing to do with probes. It was made from Earth's surface. On June 22, 1978, it was found that Pluto, the most distant of the planets, has a satellite. It was named Charon (KEHR-on) after the ferryman who carried the shades of the dead across the River Styx and into Pluto's kingdom of Hades.

The satellite turned out to be surprisingly large; it seems to be about 1,300 kilometers in diameter.

The diameter of Pluto itself has been a matter of dispute since it was discovered in 1930. Before its discovery it was assumed that another outer planet would be a gas giant like the rest. Once discovered, Pluto was found to be surprisingly dim, so that it had to be smaller than expected. With every new assessment its size shrank. It began to seem no larger than Earth, then no larger than Mars.

Once Charon was discovered, the total mass of Pluto and Charon could be calculated from the period of revolution and the distance between them. From the comparative brightness of the two, the separate masses could be determined and, assuming the density to be that of ice, the diameter could be estimated. It turned out that Pluto has a diameter of about 3,000 kilometers, and is therefore just a bit smaller than Europa, the smallest of the large satellites.

Now for a sidelight—

Each planet is more voluminous than all its satellite material. Mercury and Venus are infinitely more voluminous than their satellites, since they have none, while Mars is 15,500,000 times as voluminous as its two satellites taken together. Jupiter is about 8,500 times as voluminous as all its satellites put together, and Saturn is about 8,800 times as voluminous as its satellite system.

Uranus does a little better, having no large satellites, and is about 10,000 times as voluminous as its satellite system. Neptune, with a large satellite, and being itself the smallest of the four large gas giants, does somewhat worse and is only about 1,350 times as large as its two satellites taken together.

We might conclude from this record of seven of the nine

planets that it is a cosmic law that any planet must be at least 1,000 times as voluminous as all its satellites put together.

But then we come to Earth, and behold!—Earth is only 50 times as voluminous as the Moon.

Earth people are rightly proud of this, and owning such a large satellite has been very useful, considering how it has helped in the intellectual advance of the human species (see "The Triumph of the Moon" in *The Tragedy of the Moon* (Doubleday, 1973). We consider ourselves to be the nearest approach to a double planet in the solar system, and I even published a book once about the Earth-Moon system, which I called *The Double Planet* (Abelard-Schuman, 1960).

Well, here's a case where I have to update even the title of a book, for Pluto is only *twelve* times as voluminous as its satellite, and Pluto-Charon is a far closer approach to a double planet than Earth-Moon is.

Too bad.

6
ARM OF THE GIANT

In addition to being a prolific writer, I'm a prolific speaker, averaging very nearly a speech a week these days. One difference in my two careers, however, is that whereas there are professional literary critics, there are no professional oratory critics.

Believe me, I don't complain about this lack. I share with all other writers I know (living and dead) a low opinion of professional critics and I ask for no new variety of the species. As far as my speeches are concerned, I am delighted to accept the applause and ovations at face value; I am pleased to have people come up to say pleasant things; and (best indication of all) I am gratified to have the person who persuaded me to come hand over the check with a big grin on his or her face.

I don't need the addition of someone making a living out of explaining where I fell short.—And yet, something of the sort sometimes shows up unexpectedly. (Or, as some unsung philosopher once said, "You can't win them all.")

ARM OF THE GIANT

I was asked to give an evening talk a few weeks ago to a convention of the American Psychiatric Association. When I asked what on earth I could tell a couple of thousand psychiatrists, considering that I knew nothing about psychiatry, I was told expansively, "Anything you wish."

So I talked about robots and their effect on society and what the future of robotics might hold for us. I introduced the subject by telling them in humorous detail how I came to write my robot stories and recited the Three Laws of Robotics and was, as I usually am, very self-assured and unhumble.

The talk gave every sign of being a huge success and I was pleased. My dear wife Janet, however (who is herself a psychiatrist), had taken a seat in the very last row in order to be less conspicuous, and she now seemed a bit depressed. I could see that, so I asked about it and she explained.

After I had been speaking for a while (Janet told me) a woman seated near her began talking loudly to her neighbor. Janet attracted her attention and asked her very politely to keep her voice down.

At which the woman said with contempt, "Why? Don't tell me you find him interesting? It's nothing but narcissistic nonsense."

Naturally I laughed, and told Janet to forget it. I have never expected to please everyone.

Then, too, I don't know if the woman was herself a psychiatrist or had just wandered in off the street, but she was certainly not using "narcissistic" in its psychiatric sense. She used it in its casual, everyday meaning of "abnormally interested in one's self to the disregard of others," and to grasp the fact that I am narcissistic in that sense is no great discovery.

In fact, just about everyone is narcissistic in that sense, usually with far less excuse than I can manufacture. For instance, my critic was being rather nastily narcissistic by deliberately expressing her displeasure at me in a way that would disturb others who might, conceivably, have been interested in my talk.

We don't even have to confine ourselves to individuals. The human species as a whole is unbelievably narcissistic and, by and large, considers itself to be the whole reason for the existence of the universe. Their interest in anything else at all is confined almost entirely to objects that impinge upon them and in direct proportion to the extent of that impingement.

For instance, it is estimated that there are 10^{22} stars in the known universe and yet humanity generally fixes its attention on just *one* of them (the Sun) to the nearly total exclusion of the others, just because it happens to be the closest to ourselves.

Just to show you what I mean, we will all agree at once that the Sun is the star which is by far the largest in apparent size. After all, it is the only star that appears in the form of a disc rather than a mere point of light. Very well, then, but which is the *second*-largest star in apparent size? How many people know? Or care?

To discourage narcissism, therefore, I will now take up the question of the second-largest star in apparent size.

The constellation of Orion is generally considered the most beautiful in our northern sky because it is so large, so interesting in shape, and so rich in bright stars. The name of the constellation dates back to the Greeks who had a number of myths about a giant hunter named Orion. He was loved by Artemis, the goddess of the hunt, but she was maneuvered into killing him by her brother, Apollo. In sorrow and contrition, she translated him into the sky as the constellation.

The giant hunter in the sky is usually pictured as holding a shield in his left arm to ward off the charge of Taurus (the Bull), while his right arm holds his club up high, ready to bring it down upon the furious animal. A bright star marks each of these arms. Farther down, a bright star makes each of his legs. Between the two is a horizontal line of three bright stars which marks out his waist (Orion's belt).

The brightest of the stars in Orion is one that is distinctly red in color and that shines in his right arm. Its astronomical name is Alpha Orionis.

In the early Middle Ages the triumphant Arabs took over Greek science, including the Greek picture of the heavens, and they too saw the constellation of Orion in the form of a giant hunter. The Arabs had the sensible habit of naming stars from their position in a constellation, so they named Alpha Orionis *yad al-yawza,* meaning "arm of the giant." For some reason, some European translator of an Arabic text transliterated the Arabic symbols as *bayt al-yawza* ("house of the giant," which makes no sense) and spelled it, in Latin letters, as Betelgeuse—which remains its name to this day.

In my youth I was under the impression that it was a French word, and I tried to pronounce it as such. I had nothing but contempt for anyone so illiterate as to pronounce it as though it were spelled "beetle-juice." Imagine my embarrassment, then, when I found that the dictionary pronunciation of the star is indeed quite close to beetle-juice, and you might as well say that.

Well, as it happens, Betelgeuse is better known in detail than any star but our Sun.

Why?

Consider that (all other things being equal) a nearby star is more likely to be understood in some detail than a distant one is—just as the Moon was known in surface detail long before Mars was.

Then again (all other things being equal), a large star is more likely to be understood in some detail than a small one is—just as Jupiter's surface was known in more detail, until recently, than the much smaller though nearer Phobos was.

If we want to know the details of some star other than our

Sun, we would do well, then, to choose one that is both large and nearby.

Betelgeuse is not a star that is really nearby; there may be as many as 2,500,000 stars that are closer to us. Still, considering that there may be as many as 300,000,000,000 stars in the galaxy, there are 120,000 times as many stars in our galaxy that are farther than Betelgeuse than there are stars that are nearer. We can therefore fairly say that Betelgeuse is in our stellar neighborhood.

On the other hand, we can also conclude that Betelgeuse is unusually large just by looking at it with the unaided eye. This may seem strange, since all the stars look like mere points of light not only to the unaided eye but even through the largest telescope. How, then, can we so easily tell that one point of light is larger than another point just by looking at it without instruments?

The answer is that red stars are red because their surfaces are relatively cool. Because those surfaces are cool, they have to be dim per unit area. If red stars are nevertheless very bright, that must be because they are unusually close to us, or, if that is not so, because the total surface is unusually large.

Thus the star Alpha Centauri C (Proxima Centauri) is closer to us than any other star in the sky, but that is insufficiently close even so, for it remains invisible to the unaided eye. It is red and cool, and small in addition, you see. Betelgeuse is as red as Alpha Centauri C, and it is 150 times as far away from us as Alpha Centauri C is, but Betelgeuse is not only visible to the unaided eye, it is among the dozen brightest stars in the sky. It must therefore be deduced from that fact alone as having an enormous surface.

So must have reasoned the German-born American physicist Albert Abraham Michelson (1852–1931). In 1881 Michelson had invented the interferometer, which could measure, very delicately, the way in which two beams of light would interfere with each other, the light waves of one canceling those of the other in some places and reinforcing them in others (depending on whether one wave went up while the other went down, or both went up and down together). The result was a kind of striped, light-and-dark alternation, and from the width of the stripes much could be deduced.

If a star as seen by us in the sky were a true point, with zero diameter, all the light would come in a single ray and there would be no interference whatever. If, however, a star had a finite diameter (however small), the light from one side of the star and the light from the other side would be two separate rays that would converge toward the point of observation, forming a very tiny angle. The two separate rays would interfere with each other, but would do so extremely slightly and the interference would be very difficult to detect. Naturally the larger the star, the larger the angle (though it would still be tiny) and the better the chance of detecting interference.

Michelson used a special interferometer, six meters long, that was capable of detecting particularly tiny effects. He also made use of the then new 2.5-meter (100-inch) telescope, the largest in the world. In 1920 the apparent diameter of Betelgeuse was measured. It was the first star shown by actual measurement to be more than a point of light and the news made the front page of the New York *Times*.

The apparent diameter of Betelgeuse turned out to be about 0.02 seconds of arc.

How wide is that?—If you imagined 100,000 shining dots like Betelgeuse side by side and touching, you would have a thin, bright line with a length equal to the width of the full Moon when it is nearest Earth. If you imagined 65,000,000 dots like Betelgeuse side by side and touching, you would have a thin bright line circling the sky like a gleaming equator.

What's more, if you imagined a great many shining dots, each the apparent size of Betelgeuse, and imagined them packed tightly together over the sphere of the sky, it would take about 1⅓ quadrillion of them (1,330,000,000,000,000) to convert the sky into a solid red blaze all around the Earth.

When you think of that and realize that, in actual fact, the star is spangled with only 6,000 visible stars, you will realize how empty the sky actually is, even allowing for the Sun, Moon, and six visible planets.

Betelgeuse is a variable star—that is, its brightness varies with time. What's more, there is no simple periodicity to the variation, so it is an "irregular variable." Its average brightness is of magnitude 0.85, but it brightens to 0.4 at times and sinks down to 1.3 at others.

The reason for this variability is not mysterious. The mere fact that a star is a red giant means that it is in its final stage as an extended star. Before long, it will no longer be able to support the mass of its outer layers by the energy of the fusion reactions deep within itself and the star will then collapse (with or without an explosion). The fact that Betelgeuse is flickering, so to speak, is another indication that it is close to the end. The flicker is due to turbulence and various instabilities to be expected of a star that is having trouble supplying itself with enough heat to keep itself expanded.

That being so, then there should be noticeable changes in the diameter of Betelgeuse as measured by interferometer, and there are. The apparent diameter varies from 0.016 to 0.023 seconds of arc.

In order to tell how large Betelgeuse actually is, in absolute units, from its apparent size, you must know its distance, and that's not easy. Stellar distances that are greater than 30 parsecs (100 light-years) or so are difficult to determine.

The latest (and presumably most nearly reliable) figure I have been able to find for the distance of Betelgeuse is 200 parsecs (650 light-years).

For the sphere of a star to appear to be 0.02 seconds of arc in diameter even though it is at a distance of 200 parsecs, it ought to have a real diameter of about 600,000,000 kilometers, if my calculations are correct. Betelgeuse has a diameter, then, that is on the average 430 times the diameter of the Sun. Its volume would then be 80,000,000 times that of our Sun, which means that if Betelgeuse were imagined as a hollow sphere, you could drop 80,000,000 spheres the size of the Sun into it before the large sphere was filled (assuming the small spheres were packed tightly together so that there was no space between them).

If you imagined Betelgeuse in the place of our Sun, its surface would be located nearly at the orbit of Mars. Earth's position would be seven tenths of the way from Betelgeuse's center to its surface.

We can now get a more dramatic picture of its pulsation. When Betelgeuse expands to its maximum, its diameter would increase to about 725,000,000 kilometers, or just about 500 times that of the Sun. At its minimum, it would be a mere 500,000,000 kilometers, or 360 times that of the Sun.

At full expansion, the surface of Betelgeuse, if it were imagined in the place of our Sun, would be well into the asteroid belt. It is three times as voluminous at maximum as at minimum. If it is pictured as breathing hard because it is near the end of its race as an extended star, it is breathing *very* hard.

Granted that Betelgeuse is a giant star in reality (it belongs, in fact, to a class of star we call "red giant"), how does it compare in *apparent* size with other stars that may be smaller—but closer.

For instance, I have already said that Alpha Centauri C is the star closest to ourselves. It is part of a group of three stars, the largest of which is Alpha Centauri A. Alpha Centauri A is almost exactly the size of our Sun, and at its distance of 1.35 parsecs (1/150 that of Betelgeuse) its apparent diameter would be only about 0.0035 seconds of arc, less than one fifth that of Betelgeuse. For all Alpha Centauri A is so close, its puny size cannot show as large as distant, giant Betelgeuse.

Sirius is larger than Alpha Centauri A, but it is also farther and its apparent diameter is only about 0.0032 seconds of arc. Arcturus is 32,000,000 kilometers in diameter (23 times that of the Sun), but it is 11 parsecs away and its apparent diameter is 0.0095 seconds of arc, while Aldebaran is 50,000,000 kilometers in diameter (36 times that of the Sun) but is 16 parsecs away, so that its apparent diameter is 0.011, just about half that of Betelgeuse.

There just isn't any star that is sufficiently large or sufficiently close (or both) to rival Betelgeuse. The one that comes closest is another red giant, Antares, in the constellation of Scorpio. It is at a distance of 130 parsecs, so it is closer than Betelgeuse, but it is slightly dimmer than Betelgeuse even so, despite the advantage of nearness, and must therefore be appreciably smaller.

As it happens, Antares, at its distance, would have an apparent diameter of 0.002 seconds of arc, which is equal to the average value for Betelgeuse, but Antares does not pulsate appreciably. It is therefore smaller in apparent size than Betelgeuse is at maximum.

In short, of all the stars Betelgeuse is second to the Sun in apparent size.

*   *   *

Betelgeuse has a surface temperature of 3,200° K° as compared to our Sun's surface temperature of 5,700° K. Betelgeuse is red-hot, while our Sun is white-hot.

If the Sun's surface temperature were suddenly to shrink to 3,200° K., then aside from the fact that its light would redden, it would deliver a total illumination only about 1/43 as intense as it does now.

Betelgeuse has 185,000 times as much surface as the Sun has, however, so that even though each Sun-sized portion delivers only 1/43 the illumination of our Sun, the entire star blazes with a light 4,300 times that of the Sun.

Astronomers make use of the term "absolute magnitude" to represent the brightness a star would display if it were exactly 10 parsecs from Earth. If we were to view our Sun from a distance of 10 parsecs, it would have an absolute magnitude of 4.7, which would make it a rather dim and unspectacular star.

Betelgeuse, on the other hand, if moved toward us to a distance of 10 parsecs, would blaze with an absolute magnitude of –5.9. It would shine, ruddily, with a brightness 4 1/3 times that of Venus at its brightest.

It would then have an apparent diameter of 0.4 seconds of arc, which would be enormous for a star (other than our Sun), but it would still appear to be merely a point of light. After all, the planet Jupiter has an apparent diameter of 50 seconds of arc, and still looks like a mere point of light to the unaided eye.

Despite the enormous size and brilliance of Betelgeuse, in some ways it is not quite the giant it appears. What about its mass, for instance—the quantity of matter it contains?

It is more massive than the Sun certainly, but not enormously more massive. In fact, it is estimated that it is 16 times as massive as the Sun. *Only* 16 times.

This mass of 16 Suns is spread out over a volume that is, on the average, 80,000,000 times that of the Sun. The average density of Betelgeuse must therefore be $16/80,000,000$ or $1/5,000,000$ that of the Sun.

This is smaller than you might expect, for it amounts to about $1/4,500$ of the density of the air you are now breathing. When Betelgeuse is at its fullest expansion, the quantity of matter it contains is stretched over an even larger volume, and its average density is then $1/7,000$ that of air.

If we were to suck out all but $1/4,500$ of the air in some container, we would be justified in speaking of the result as a vacuum. It wouldn't be an absolute vacuum, or even a very hard one, but it would be vacuum enough for the practical everyday sense of the word. It would be rather natural, then, to view Betelgeuse (or any red giant) as a kind of red-hot vacuum.

Still, Betelgeuse (or any star) is not evenly dense all the way through. A star is least dense at its surface and that density rises, more or less steadily, as one penetrates below that surface, and is highest of course at the center. The temperature also rises to a peak at the center.

A star begins as a ball of hydrogen, chiefly, and it is at the center, where the temperature and density are highest, that nuclei smash together hard enough to fuse. It is at the center, then, that hydrogen is fused to helium and energy is developed. The helium accumulates, forming a helium core that grows steadily as fusion continues.

Hydrogen fusion continues to take place just outside the helium core where the hydrogen is at the highest temperature and density; and the helium core, as it grows, becomes hotter and denser itself. Eventually, after millions or even billions of years, the temperature and density in the helium core become great enough to force even the stable helium nuclei to fuse further into carbon and oxygen nuclei. (Carbon nuclei are composed of three helium nuclei; oxygen nuclei of four.)

The new surge of heat developed by the onset of the helium fusion causes the star (which, all during hydrogen fusion, has remained relatively unchanged in appearance) to expand so that the surface cools. The star, in other words, which has been till then a white-hot, relatively small object, suddenly expands into a red giant as a new core of carbon and oxygen forms and grows at the center.

That, then, is the situation with Betelgeuse. At its center is a carbon-oxygen core that is at a temperature of $100,000,000°$ K. (as compared with $15,000,000°$ K. at the center of the Sun). This is still not hot enough to cause the carbon and oxygen to fuse to even more complicated nuclei.

This core (as best astronomers can tell from computer calculations based on what they know of nuclear-reaction theory) is perhaps twice the diameter of the Earth and has a density something like 50,000 grams per cubic centimeters, or more than 2,000 times that of platinum on Earth. Betelgeuse is certainly not a red-hot vacuum all the way through.

Perhaps $\frac{1}{50}$ of the total mass of Betelgeuse is packed into that small core. Around the core is a helium shell, perhaps ten times the volume of the core, that holds another $\frac{1}{50}$ of the total mass. And outside the helium shell are the rarefied outer regions that are still to a large extent hydrogen. Helium continues to fuse at the surface of the carbon-oxygen core, and hydrogen continues to fuse at the boundary of the helium shell.

Hydrogen at the bottom of the rather rare hydrogenous outer region of Betelgeuse cannot fuse at the enormous speed with which it would have fused at the center. Helium, fusing under denser and hotter conditions, produces much less energy per nucleus. The two fusions together can barely produce enough heat, therefore, to keep Betelgeuse in its state of enormous distention. Every once in a while there is a shortfall, apparently, and the star starts to contract. The contraction compresses the hydrogen and helium and speeds the fusion, so that the star expands again.

As time goes on, further fusion reactions take place at the core, each one producing less energy per nucleus than the one before, so that the situation gets steadily worse. Eventually, when iron nuclei form at the core, there is no way for any further energy-producing fusion to take place there, and periodic contractions become more and more extreme. Finally, there is one last failure and the star collapses altogether and permanently.

The sudden collapse will compress what fusible material still remains and much of it will go off all at once to produce an explosion. The more massive the star, the more sudden the collapse and the more catastrophic the explosion.

A star the size of the Sun will collapse and fizzle, blowing a small portion of its outermost layer into space. This will form a spherical shell of gas about the collapsed star. Seen from afar, the shell will look like a smoke ring and the result is a planetary nebula. The collapsed star at the center will be very small and dense—a white dwarf.

A star considerably larger than the Sun—like Betelgeuse— will explode violently enough to be a supernova. The compressed remnant will collapse beyond the white-dwarf stage and will become a neutron star or even, perhaps, a black hole.

This is undoubtedly the fate to be expected of Betelgeuse in the comparatively near future, but to an astronomer the "near future" could mean 100,000 years, so don't wait up nights for it. There is at least one other star that seems likely to beat Betelgeuse to the punch (see "Ready and Waiting" in *X Stands for Unknown*, Doubleday, 1984) and even there, it may be a few thousand years before it explodes.

Even barring a supernova, however, there is more of interest to say about Betelgeuse in the next chapter.

7

THE WORLD OF THE RED SUN

When I was a little younger than I am now, and was in junior high school, I used to read the science fiction magazines that were to be found on the magazine stand in my father's candy store.

Those stories that particularly appealed to me I would retell to an absorbed group of classmates during lunch hour, and the most successful example of these secondhand narrations was that of a story I loved called "The World of the Red Sun," which appeared in the December 1931 issue of *Wonder Stories*.

I took no note of the author's name at the time, for he was by no means well known. The story was, in fact, the first he published.

It was only many years afterward, during which time I had come to be a correspondent and good friend of the famous s.f. writer Clifford D. Simak that, obtaining Donald Day's invaluable index of science fiction stories from 1926 to 1950, I looked up "The World of the Red Sun" and discovered it was the maiden effort of none other than Cliff. (And now, over half a century later, he is still active, still turning out crackerjack material, and has been voted a Grand Master by the Science Fiction Writers of America.)

It has always been a source of infinite satisfaction to me that, when a mere preteen child, I recognized greatness in an author's first story.

You can imagine, then, the pleasure with which I came to realize, as I planned this essay, that the most logical title for it would be the one Cliff gave his first story. I am using that title, therefore, in homage to an old friend.

Cliff's story was a tale of time travel, and the Red Sun he spoke of was our own Sun in the far future. My Red Sun, however, is the star I dealt with in considerable detail in the preceding chapter—Betelgeuse, the red giant.

The question is this: If we consider Betelgeuse as the Red Sun, can there be a world circling it? By that I don't mean a planet, plain and simple—but a particular planet, one that is Earthlike in character and has intelligent life upon it. The life doesn't have to be humanoid, of course, but it should be life as we know it—nucleic acid and protein, built up in a watery background.

Let's see, then . . . Suppose we have an Earthlike planet to begin with (and I have a strong temptation to use "terroid" as a synonym for "Earthlike" even though I don't think this has ever been done).

A terroid planet can't be too close to a star, or its ocean will boil; it can't be too far from a star or its ocean will freeze, and, in either case, terroid life would be impossible.

Since Betelgeuse is, on the average, a star with 430 times the diameter of our Sun, we know that our terroid planet will have to be much farther from it than Earth is from the Sun. As a first approximation, let's place the planet at such a distance that Betelgeuse will have the same apparent size in its sky as our Sun has in Earth's sky.

In that case, the planet would have to be at an average distance of 63,500,000,000 kilometers from Betelgeuse ($\frac{1}{15}$ of a light-year), or over ten times the average distance of Pluto from our Sun.

If there were a planet at such a distance from our Sun, it would complete one revolution about the Sun in about 8,742 Earth-years.

Betelgeuse, however, is about 16 times as massive as our Sun, so it would whip such a distant planet faster along its orbit than our Sun could. The planet we are imagining for Betelgeuse would complete one revolution about its distant but massive star in but one fourth the time it would have taken to circle our Sun. Its period of revolution about Betelgeuse would be 2,185 Earth-years.

Does it matter that the planet's period of revolution would be over two millennia long?

Suppose it is just like our Earth. Suppose its orbit is circular, that it rotates about its axis in 24 hours, that it has an axial tilt like ours, and so on. It would then have seasons like ours, but each season would be five centuries long. *Too* long, of course. The polar regions would have centuries of continuous light and then centuries of continuous darkness.

Well, then, imagine the axis upright: 12 hours of daylight and 12 hours of night everywhere. The polar regions would undoubtedly have permanent ice caps that might extend far into the temperate zones without a hot summer to do a lot of melting, but the tropic regions would be fine. It would seem we are all set, but—

No good!

Betelgeuse is red, not white. Its surface temperature is 3,200 K., and not the 5,800 K. of our Sun. Size for size, Betelgeuse's surface would deliver only $\frac{1}{43}$ the total light and heat that our Sun does. It might look just as large as the Sun in its planet's sky, but it would be a cold sun by our standards, so the planet's ocean would freeze and terroid life would be impossible.

In that case, let's move the terroid planet inward. (The imagination is a powerful tool.) Forget about having Betelgeuse appear the size of our Sun, but let it expand as the planet moves nearer until the increased area of its surface makes up for its coolness.

We must move inward until the apparent area of Betelgeuse in the terroid sky is 43 times that of our Sun in Earth's sky, and Betelgeuse's apparent diameter, therefore, 6½ times that of our Sun. That means we must imagine the terroid planet at an average distance of 9,680,000,000 kilometers from Betelgeuse, or only 1.6 times the distance of Pluto from our Sun.

At this distance, Betelgeuse would seem to be about 3.5 degrees in diameter and it would seem bloated indeed to our Sun-accustomed eyes, but it would deliver only the proper amount of light and heat.

To be sure, the light would be different in quality. It would be reddish in color and, to our eyes, less satisfactory. However, the living organisms on Betelgeuse's planet would presumably be adapted to the star's radiation range. Their eyes would be more sensitive to red than ours are and would respond some distance into the infrared (and presumably be unaffected by the shortwave light that would be present in only small quantities after all in Betelgeuse light.) To terroid eyes, Betelgeuse's light would appear white, and organisms possessing those eyes would be perfectly contented.

What's more, the period of revolution would be shorter under these conditions, and would be only 130 Earth-years long. A slight axial tipping would be bearable and might reduce the polar ice cover appreciably.

Sounds great, doesn't it?—No, it's no good!

Our Sun is a stable star, unchanging in size and in the amount of radiation it delivers. Sure, it is spottier at some times than at others, and in recent years there have been some observations that have led astronomers to think that its size is changing very slightly with time, but these changes are trivial in comparison with the case of Betelgeuse, which, as I pointed out in the last chapter, pulsates markedly—and irregularly.

I said that Betelgeuse is 430 times the diameter of the Sun, but that is *on the average*. It can swell up till it is 500 times the diameter of the Sun (sometimes even more) or shrink till it is only 360 times the diameter of the Sun (sometimes even less).

The planet we are imagining to be circling Betelgeuse would therefore see the star with an apparent diameter of 3.5 degrees only on the average. That diameter would vary from 4.2 degrees to 2.9 degrees. At maximum diameter, Betelgeuse's apparent area in the sky would be twice what it was at minimum diameter, and it would deliver twice the radiation at maximum as at minimum.

This means that our imaginary planet is going to suffer enormously hot periods of time and enormously cold ones even if its orbit about Betelgeuse is circular and its axis is upright. I suspect, in fact, that the temperature variations on the planet would be too great for life as we know it to develop.

But does its orbit have to be circular? Might we not imagine a rather elliptical orbit so arranged that the planet approaches Betelgeuse just at the time when the star is contracting and delivering less light and heat, then moves away from Betelgeuse just as it is expanding and delivering more?

It would be asking much of coincidence to suppose that the planet rushes in and skids out again in just the proper synchronization to keep its temperature fairly steady, but I wouldn't hesitate to imagine it just because it's unlikely.

The trouble is that it isn't just unlikely; it's *impossible!*

I said that the planet would be circling Betelgeuse in 130 years. No matter how highly elliptical the orbit might be, the period of revolution would still be 130 years if the average distance from Betelgeuse remained 9,680,000,000 kilometers. That means it would be relatively close to Betelgeuse for something less than 65 years, and relatively far from it for something more than 65 years. The reason for this is that the planet would move at a faster than average orbital speed when closer to Betelgeuse and at a slower than average one when farther. The more highly elliptical the orbit, the more unbalanced the times it would spend near and far.

There is no way of making this situation match the expansion and contraction of Betelgeuse unless the star expanded and contracted with a 130-year period and with the expanded part of the cycle somewhat longer than the contracted part.

The pulsation period of Betelgeuse isn't even close. It takes about 50 days for Betelgeuse to expand from minimum size to maximum, and about 100 to 150 days for it to contract from maximum to minimum again. In one orbital period of the planet about Betelgeuse, then, the star would expand and contract about 270 times. To balance this, you would have to wave the planet in and out, in varying periods and to a varying extent, in order to match exactly the unpredictable variations in the rate and extent of Betelgeuse's expansion and contraction.

Apparently the irregularity of Betelgeuse traces back to the fact that it is turbulent and "boiling." Hot bubbles of helium from the interior periodically rise to the surface and produce enormous hot spots that cause the star to expand. The variables involved are too many to allow much in the way of regularity.

You might argue, of course, that Earth has a great deal of variation in its weather, too, and yet life exists here.

But then, Earth's temperature variations as a whole are far less than those that Betelgeuse's planet would be compelled to endure and, furthermore, there are large regions on Earth where the temperature is quite equable over long periods of time. It is difficult to see how this would be true of Betelgeuse's planet, too.

Betelgeuse is enormously unstable in other ways as well. It shows signs of possessing colossal prominences and of being the source of a huge stellar wind. This all argues that Betelgeuse is not going to remain in its present form for long, as compared with ordinary stars like our Sun, which can continue relatively unchanged for billions of years.

Compare Betelgeuse's wind to that of the Sun. The Sun is constantly losing mass, as streams of particles (chiefly protons— the nuclei of hydrogen atoms, which make up the bulk of the Sun's substance) speed outward in all directions. About a million metric tons of matter are lost to the Sun each *second* through this solar wind, but Betelgeuse loses matter at a billion times that rate.

If Betelgeuse were to continue to lose mass through its stellar wind at the present rate, it would be entirely gone in 16,000,000 years. The chances are, though, that long before this Betelgeuse would either have blown off enough matter to be converted into a condensed star surrounded by a planetary nebula, or would have blown up in a supernova. I suspect that a large red giant can only remain in that state about 2,000,000 years.

That might seem ample time to you, considering that human civilization has lasted less than 10,000 years. A period of 2,000,000 years is two hundred times that long.

No good! We're not talking about the development of civilization, but the development of life. Life appeared on Earth perhaps 3,500,000,000 years ago, and multicellular life perhaps 800,000,000 years ago, and land life only 400,000,000 years ago. It took more than two and a half billion years just to pass beyond the one-celled stage, and that is over a thousand times as long as the lifetime of a red giant.

You might say that evolution just happened to be extremely slow on Earth and that it might be faster on Betelgeuse's planet.

Well, we can't tell whether the rate of evolution on Earth is, or is not, typical of life in the universe generally because Earth's life is the only sample of the phenomenon we know. Yet, from what we know of evolution it seems hard to suppose that it can be anything but a very slow process. It is difficult to believe that intelligent life could evolve during the brief existence of a red giant.

In that case, let's remember that Betelgeuse wasn't a red giant to begin with. Before it was a red giant, it was on the main sequence. That is, it was a stable star like the Sun, subsisting by hydrogen fusion at the core. It was then a relatively small star—more massive than the Sun and therefore somewhat larger, brighter, and hotter, but no giant.

Why, then, should we suppose that life had to begin while Betelgeuse was a red giant? Would it not make sense to suppose that life began when it was on the main sequence, and that life developed to intelligence, and to high technology, during that period?

Then, when Betelgeuse came to the end of its main-sequence life and began to evolve into a red giant, the intelligent inhabitants of the original terroid planet (which would, of course, be circling Betelgeuse at a greater distance than Earth is from the Sun, since Betelgeuse was the hotter star—but not at a terribly greater distance) would have the space-traveling capacity to move farther outward. The movement would be in stages because although the evolution to the red-giant stage is rapid as compared to change during the main sequence, it is still quite slow on the human life scale.

Thus when our Sun begins evolving to the red-giant stage, human beings (or our evolved descendants), if still in existence, might slowly move out to Mars; then, hundreds of thousands of years later, to Europa; then, a million years later, to Titan, and so on. Betelgeuse, being more massive, would evolve more rapidly than the Sun would, but there would still be no hurry.

Therefore, the distant planet of Betelgeuse's red-giant stage would not carry intelligent life that had developed there, but life that had *migrated* from some inner planet that had been physically vaporized and absorbed by Betelgeuse as that star expanded.

No good!

In our solar system, the worlds relatively close to the Sun are essentially rock, with or without a metallic core, and could conceivably support long-term human life, either naturally (as Earth does) or after considerable technological modification as the Moon or Mars might.

The worlds beyond the asteroid belt, which will survive the solar red giant, are, however, of fundamentally different composition. The large worlds are chiefly gaseous, while the small worlds are chiefly icy. Such worlds do not offer much hope as long-term refuges. The gaseous ones are too entirely alien. The icy ones don't have the rocky and metallic elements we need.

Of course, the solar red giant may conceivably heat up Jupiter to the point where much of it will be dispersed, and we might dream that a rocky core will be exposed that would be a fresh new Earth. Unfortunately, we're not sure that there is a rocky core at all, or how large it might be—or, for that matter, whether even a heated Jupiter might not cling together more or less intact, thanks to its large gravitational field.

Of the large Jovian satellites, Ganymede and Callisto are icy, and at red-giant time may melt and disperse. Io, to be sure, is rocky, but lacks water. Callisto is rocky, and has a world-girdling surface ocean (now frozen, at least on top). The red giant may melt and vaporize the ocean, which might thus be lost into outer space.

Beyond Jupiter, eveything should remain intact, but the worlds are not really inviting.

There's every reason to think that this general pattern—rocky worlds near a star, and gaseous or icy worlds far from a star—is general in planetary systems. We might expect, then, that it is a rule that life begins relatively close to a star and that in red-giant time, retreat to the outer regions would involve such extensive terraformation as to be prohibitive.

But are we not putting shortsighted limits to the possible advance of technology? Terraforming may be very simple to a technologically advanced species. Considering the rate of technological advance in the past hundred years (from the unpowered glider to rocket probes taking close-up pictures of Saturn's rings), what might we not expect of ourselves in another hundred years, let alone a thousand?

And who says we have to be satisfied, as refugees, with whatever world happens to exist in the outer reaches of a planetary system? They are only resource accumulations.

We can picture humanity, as the time for the solar red giant approaches, to be living in artificial space settlements, each as comfortable and pleasant as Earth's surface and much more secure. There might never be any thought of returning to any world. One would merely move the settlements farther from the Sun, little by little, year by year, keeping pace with the upward creep of solar radiation intensity.

We might even picture humanity as saving worlds from solar destruction, pushing them farther from the Sun every once in a while, in order to keep them as resources.

Therefore, we might picture the life that originally developed relatively near Betelgeuse in its main-sequence days, as now living in large settlements nearly ten billion kilometers from the star, with rescued asteroids and satellites also in orbit. We might even suppose the inhabitants to have methods for flattening out the differences in radiation received as Betelgeuse expands and contracts. They could shield the settlements and deflect most of the radiation as Betelgeuse heats up, and could gather and concentrate radiation as it cools down.

No good!

All of this depends on whether life could really have begun and developed in the Betelgeuse planetary system while that star was still on the main sequence.

Let's consider our Sun, for instance, and in doing so let's not deal in billions of years. It is hard to grasp such enormous periods of time. Let us define ''6 long-years,'' instead, as equal to 1,000,000,000 ordinary years. On this scale ''1 long-second'' is equal to 31 years.

Using this ''long-standard,'' the solar system would condense out of a primordial swirl of dust and gas in about 7 long-months and enter its existence on the main sequence. It would remain on the main sequence for about 72 long-years (about the average lifetime of a human being, which is why I chose this particular scale) then flash through the red-giant stage in no more than 4 *long-days*, and collapse to a white dwarf, in which state it will remain indefinitely, slowly cooling off.

If we look a little closer at the main-sequence portion of the Sun's lifetime, and do so in long-years, here are the results.

The planets and other cold bodies of the solar system took on their present shape only slowly as they collected the debris in their orbits. The bombardment of this debris has left its mark in the form of the meteoric craters that scar every world where they are not eroded or obscured by such factors as air, water, volcanic lava, living activity, and so on. It was not until the Sun was 3 long-years old that this bombardment was essentially over and Earth and the other worlds were in more or less their present shape.

When the Sun was 6 long-years old, the first traces of molecules, complicated enough to be considered as living, appeared on the Earth.

When the Sun was 21 long-years old, the first multicellular life formed, and at 24 long-years of age the fossil record begins. The Sun was a little past 25 long-years old when life crawled out on land, and it is now a little past 27½ long-years old. By the time it is 60 long-years old, it may be a little too hot for Earth to be comfortable, and human beings or their evolved descendants (if still around) may begin the retreat outward. By the time it is 72 long-years old, our Sun will be a red giant, though not as large as Betelgeuse is now.

As it happens, not all stars remain on the main sequence an equal length of time. In general, the more massive a star is, the greater its nuclear fuel supply. The more massive it is, however, the more rapidly it must consume that fuel supply if it is to generate enough heat and radiation pressure to keep itself from collapsing under the pull of its greater mass.

The rate of fuel expenditure rises more rapidly than the fuel supply does, as the mass goes up. It follows that the more massive a star is, the shorter its time on the main sequence and the more rapidly it reaches the red-giant stage.

THE WORLD OF THE RED SUN

Consider the red dwarfs, which make up three quarters of all the stars. These are relatively small stars with masses from ⅕ to ½ that of the Sun, just massive enough to produce internal pressures capable of igniting nuclear reactions. They dribble out their relatively small fuel supply so slowly that they remain on the main sequence for lengths of from 450 long-years up to as much as 1,200 long-years.

These are enormous life-spans when you think that the universe itself is thought to be not more than 90 long-years old at the present time. This means that every red dwarf in existence is still on the main sequence. Not one has yet had time to reach the red-giant stage.

On the other hand, stars that are more massive than the Sun have a shorter stay on the main sequence. Procyon, for instance, which is about 1.5 times as massive as the Sun, will remain on the main sequence for a total of 24 long-years. Sirius, with a mass of more than 2.5 times that of the Sun, will be on the main sequence for only 3 long-years. (I view this sort of thing again, in a different way, in the final chapter of this book.)

And what about Betelgeuse, which is 16 times the mass of the Sun? Well, it remains on the main sequence for about 3 *long-weeks*. Compare this with the 6 long-years (a period over a hundred times as long) that elapsed before the first trace of life came into being on Earth.

Even if we grant that our solar system was phenomenally slow in developing life, it is difficult to imagine that life could develop in less than a hundredth the time.

And it isn't just the first traces of life we are interested in. We expect life to evolve *slowly* into more and more complicated forms until some species with enough intelligence to develop an advanced technology can come into being. It took Earth 27 long-years to do it. Could the Betelgeuse planet have done it in 3 long-weeks, not much more than ⅟₅₀₀th that period?

It simply seems beyond any possibility that life could have developed on any planet circling Betelgeuse, or that there could be any home-grown life there now. (I say "home-grown" because I don't want to exclude the possibility that some beings with advanced technology, who may have originated in some other stellar system, have established a scientific observatory in the outer reaches of the Betelgeuse system in order to study a red giant at close hand. If such a station has life forms aboard, they had better be gone and a light-year away on the day on which Betelgeuse explodes.)

So there is no World of the Red Sun in the Betelgeuse sense, alas, and we can't expect terroid life to originate about any star appreciably more massive than our Sun. Stars that are appreciably less massive than our Sun are excluded for other reasons.

That leaves us only with stars reasonably close to the mass of our Sun as suitable for terroid life development. Fortunately such stars make up ten percent of the total, and that leaves us considerable leeway.

8
LOVE MAKES THE WORLD GO ROUND!

One thought leads to another and I am accustomed to letting my mind wander. For instance, something I thought of recently made me wonder about the phrase "It's love that makes the world go round!"

What this means to most people is that love is so exalting an emotion that to experience it makes one feel that the whole world is new and wonderful, while to lose it makes the Sun itself seem to lose its brightness and the world to cease its turning.—That sort of nonsense.

And who said it first?

I turned to my reference library and found, to my considerable astonishment, that the first use in English literature was in 1865, when the Ugly Duchess says in Lewis Carroll's *Alice in Wonderland* "And the moral of that is 'Oh, 'tis love, 'tis love, that makes the world go round!' "

In the same year, it appeared (with one " 'tis love" extra) in Charles Dickens's *Our Mutual Friend*. Independent invention seems unlikely, so the sentiment must have had an earlier existence as a folk saying and, sure enough, there is a line from a French popular song of about 1700 to the effect that *C'est l'amour, l'amour, qui fait le monde à la ronde*, which translates into the Duchess's remark.

Going back further, we come to the last line in Dante's *The Divine Comedy*, which contains the phrase *L'amor che move il sole e l'altre stelle* ("Love that moves the sun and the other stars"). This refers to general motion rather than merely rotation about an axis, but it will do. And here, you see, we do not mean by "love" that sense of human romantic affection that most of us naturally think of when the word is used. Rather, Dante is referring to that attribute of God which shows its concern for humanity and keeps the universe in operation for our good and our comfort.

This in turn may have been, at least in part, inspired by an old Latin proverb dating, I suppose, from Roman times: *Amor mundum fecit* ("Love made the world").

And from here we go back to the mystical cosmogonies of the Greeks. According to what we know of the doctrines embodied in the Orphic mysteries, the universe began when night (i.e., primeval chaos) formed an egg, out of which hatched Eros (divine love), and it was this divine love that created the Earth, sky, Sun and Moon, and set it all in motion.

Metaphysically this "divine love," whether pagan or Judeo-Christian, may evidence itself in the material universe as an inexorable attraction that all objects would have for each other. There is indeed such an inexorable attraction that binds the universe together, and scientists now call it "the gravitational interaction."

What we may all really be saying, then, is "Oh, 'tis gravity, 'tis gravity, that makes the world go round," and that's not such a bad idea, perhaps.

And what started this line of thought? Well—

LOVES MAKES THE WORLD GO ROUND!

* * *

In May of 1977 an essay of mine was published called "Twinkle, Twinkle, Microwaves," in which I recounted the story of the discovery of pulsars—tiny, rapidly spinning neutron stars. These are no wider in diameter than the length of Manhattan island, and yet can contain as much mass as a full-sized star. The first pulsar to be discovered made one rotation about its axis in 1.3370209 seconds. That is a very fast rotation even for an object as small as a pulsar.

Why, then, should a pulsar spin so rapidly?

A pulsar is the remnant of a supernova—a giant star that exploded. Such an explosion would send part of the stellar mass into space in all directions in a vast, expanding cloud of gas and dust, while the central portions would collapse into an extremely dense, extremely small neutron star (or sometimes into a black hole).

The original star would have a certain amount of angular momentum—how much would depend upon its rate of rotation and upon the average distance of the matter it contained from the axis of rotation.

It is one of the fundamental laws of nature that the amount of angular momentum of a closed system cannot be changed. When a star explodes, some of the angular momentum would be carried off by the gas and dust that goes swirling outward, but a good part of it would be trapped in the collapsing central portions.

As the core of the star, with its angular momentum, collapses, the matter of which it is composed draws nearer to the axis of rotation, much nearer. From an average distance of millions of kilometers it shrinks to an average of only five kilometers. This, taken by itself, would reduce the angular momentum to nearly nothing were it not for the existence of the other factor, the rate of rotation. In order for the angular momentum to remain conserved, as it must, the vast decrease in distance from the axis must be balanced by a vast increase in the rate of rotation.

You see then what makes the pulsar go around as fast as it does. It is the collapse of the star, brought about by the inexorable pull of its own gravitation. And if we equate gravitation, mystically, with love, we find that indeed "Love makes the world go round!" (Now you see my line of thought.)

If anything, in fact, pulsars don't spin quickly enough. The enormous contraction should result in a considerable faster spin. Soon after pulsars were discovered, however, it was pointed out that there were slowing effects. Pulsars spewed out energetic radiation and particles, and the energy thus expended came at the expense of their rotational energy. As a result, the rate of rotation should slow down. Another way of putting it was that the emissions were carrying off angular momentum.

Actual measurements did show that pulsars were steadily slowing. The rotation of the first pulsar discovered is slowing at a rate that will double its period in 16,000,000 years.

From this it follows that the older a pulsar is—the longer the time lapse since the supernova explosion that formed it—the longer its period of rotation should be.

In October 1968, astronomers detected a pulsar in the Crab Nebula, a cloud of gas that formed when a supernova exploded 930 years ago. That's an extremely short time, astronomically speaking, so it was no surprise when the Crab Nebula pulsar was found to rotate considerably more rapidly than the other pulsars that had been detected. The Crab Nebula pulsar rotates on its axis in 0.033099 seconds, or 40.4 times as fast as the first discovered pulsar does. Another way of putting it is that the Crab Nebula pulsar rotates on its axis 30.2 times a second.

By 1982 more than 300 pulsars had been discovered and the Crab Nebula pulsar continued to hold the record.

This, too, was no surprise. Pulsars are very small objects and are not detectable at huge distances, so the only ones that are found are located within our own Milky Way galaxy. That means that the supernovas that formed them exploded within our own Milky Way galaxy and should very likely have been visible to the unaided eye.

Only two known supernovas have exploded in our galaxy since the Crab Nebula was formed, and those appeared in 1572 and 1604 respectively. The sites of those two supernovas have not revealed any pulsar, but then not every supernova forms a pulsar, and not every pulsar that is formed spins in a direction that would cause its streams of particles and radiation to sweep across the Earth and be detectable.

With those two recent supernovas eliminated, we can be quite sure that we will not detect *any* pulsar that is younger, and therefore faster-spinning, than the Crab Nebula pulsar. Astronomers were so certain of this that none of them wished to waste their time going to all the trouble of trying to find an ultra-fast pulsar that surely did not exist.

As it happens, astronomers have prepared listings of all the radio sources detectable in the sky. Such sources do not have to be pulsars—they can be many things. They can be clouds of turbulent gas in our own galaxy; they can be distant galaxies with catastrophic events taking place at their centers; they can be even more distant quasars.

In the *Fourth Cambridge Catalogue of Radio Emitters*, there was one such source listed as 4C21.53. It had been sitting quietly on the list since the early sixties and no one had thought anything about it. The most likely way of explaining its existence was to suppose it to be a distant galaxy, too far away to be made out visually but sufficiently active to make its radio emissions detectable.

And then, in 1972, its radio image was observed to twinkle as it passed through the solar wind that sweeps outward from our Sun. That is, the image shifted its position very slightly in a rapid and erratic manner.

Twinkling, in a more ordinary sense, is familiar to us. Light passing through our atmosphere is refracted to a tiny degree, in unpredictable directions, as it moves through atmospheric regions at different temperatures. If the beam of light is fairly thick, small bits of it may shift in one direction, and other small bits in another. These may cancel out so that the entire beam seems steady.

Thus a planet such as Mars may seem like a mere dot of light, even at its nearest approach, but it is a fat enough dot so that different portions of it twinkle differently and the effect cancels out. Mars, on the whole, then, does not twinkle.

If we observe Mars through a telescope, however, we not only enlarge the entire image but we enlarge the twinkles also. If we try to see the details of the surface, we find that the twinkling blurs those details. (That is why observing Mars from beyond the atmosphere would be such an improvement.)

The stars, however, are much tinier objects, in appearance, than the planets are. So thin is the beam of light from a star, particularly a dim star, that all of it can shift erratically as it passes through the atmosphere, and it twinkles. The twinkling itself testifies to the smallness of the star's optical image.

In the same way, when 4C21.53 twinkled as it passed through the solar wind, one had to deduce that it was a very thin beam of radiation indeed. This would not be surprising if it were indeed a distant galaxy, but it is located in the constellation of Vulpecula ("Little Fox"), fairly close to the Milky Way. This means that the beam of radio waves, if it originated from outside the galaxy, would have to travel across the galaxy's long diameter to reach our instruments. So much of the radio waves would be slightly scattered by the rarefied matter lying within our galaxy (rarefied it might be, but it is much denser than the matter between the galaxies) that no matter how thin the beam might have been to begin with it would have broadened to the point where it would not twinkle.

The mere fact of twinkling, therefore, showed that 4C21.53 was located *inside* our galaxy, and that its radio beam traveled a relatively short distance to reach us and did not have time to broaden unduly, past the stage of being able to twinkle. And if it was that close and still had a beam fine enough to twinkle, 4C21.53 must be a very small object.

Then, in 1979, there was a report that if one studied the wavelength mixture of the radio beam of 4C21.53, one found that it was very poor in the higher frequencies, poorer than were most radio sources. But pulsars were characteristically poor in the higher frequencies. Could 4C21.53 be a pulsar?

The question struck an American astronomer named Donald Backer and he began to consider the matter thoughtfully. If 4C21.53 was small enough to be a pulsar and if it had the wavelength distribution of a pulsar, and if it was therefore concluded that it was a pulsar, why didn't it pulsate?

As a pulsar rotates rapidly, it emits two streams of radio waves, one from one side of itself and one from the other side. As it rotates, first one stream, then the other passes across some given observation point. If our instruments are at that point, the radio waves are detected in pulses, with the number of pulses per second dependent on the rotation period.

If the radio waves happen to miss us altogether, as they probably do in a large majority of the cases, we detect nothing at all, but if we do detect the radio waves, we should detect the pulses also. If the pulsar is very far away, the scattering by interstellar matter could blur the pulses into a more or less steady, and weak, radio beam. If the pulsar is very old, the pulses might have weakened to indetectability. However, 4C21.53 was quite close enough (only 2,000 parsecs away) for its pulses to be distinct, and the radio beam was strong enough for pulses to be easily detected if they were there.

It occurred to Backer that there was one reasonable explanation that would clear up the mystery. Suppose that 4C21.53 were spinning very rapidly, say at least three times as rapidly as the Crab Nebula pulsar. In that case its pulses would go unnoticed, since the radio observations being made were not geared for pulses quite so rapid. He tried to publish his conjecture, but his paper was rejected as too speculative, with a suggestion that was too improbable.

Backer didn't give up. He tried to get astronomers at various facilities to attempt to spot ultrarapid pulses, but over a period of three years even when he could get people to try they came up with nothing. One of the troubles (although Backer didn't know it at the time) was that 4C21.53 was actually a conglomerate of three separate, very closely spaced radio sources, one of which was in actual fact a distant galaxy. This naturally confused matters when astronomers tried to take a very detailed look at it.

In September 1982 Backer asked the people at the Arecibo radio telescope in Puerto Rico to check 4C21.53 for a characteristic known as polarization. Pulsars show very high levels of polarization, much more so than other radio sources do. The report came back that 4C21.53 showed a 30 percent polarization, which was high even for a pulsar.

This was good news indeed, for Backer was now more convinced than ever that he had a pulsar by the tail. The people at Arecibo had even gotten occasional glimpses of possible pulses.

Backer himself went to Arecibo, where he made use of

special sophisticated instruments for seven nights. By November 12, 1982, the matter was settled: 4C21.53 was found to be a pulsar, and eventually it received the new name of PSR1937 + 214.

The new pulsar became quickly known as the Millisecond Pulsar, however, for it rotated on its axis in a little more than a thousandth of a second. To be exact, its rotation period is 0.001557806449023 seconds. This means that the pulsar is rotating on its axis 642 times per second. This is not just 3 times as fast as the Crab Nebula, as Backer had suspected might be the case, but 21.25 times as fast.

Suppose the Millisecond Pulsar is 20 kilometers in diameter. Its equatorial circumference is, then, about 62.8 kilometers. A spot on its equator would travel 642 times that distance, or 40,335 kilometers, in one second. It would be traveling at about 13.5 percent the speed of light.

A pulsar has an enormous surface gravity, but even that is barely sufficient to hold itself together against the acceleration involved in such an unheard-of rotational speed. If the Millisecond Pulsar were rotating but three times faster—roughly 2,000 times a second—it would tear itself apart.

Now comes the question: What makes the Millisecond Pulsar go around so fast?

The reasonable answer is that it spins so quickly because it is brand-new. When the Crab Nebula pulsar was detected and found to be rotating about its axis 30.2 times a second after a lifetime of 930 years, astronomers calculated backward and estimated that it might have been spinning 1,000 times per second at the time of its formation.

Well, if the Millisecond Pulsar is spinning 642 times a second now, then it ought to have been formed a mere century ago or less; and if it was, that would account for everything.

But how could it be so young? If it were so young, there would have had to be a bright supernova only 2,000 parsecs away, in the constellation of Vulpecula, a century ago or less, to mark its birth, and would not that supernova have been detected?

No such supernova was detected.

Perhaps we can make up some tortured reason to explain why such a supernova wasn't detected, but putting that to one side, there's nothing to stop the astronomers from looking at the pulsar *now*—and, of course, they have looked.

If there had been a supernova at the site of the Millisecond Pulsar in the very recent past, then there would be unmistakable signs of that explosion now. The Crab Nebula supernova that took place in A.D. 1054 left behind an expanding cloud of dust and gas that is still clearly visible now. In fact, the Crab Nebula *is* that expanding cloud.

At the site of the Millisecond Pulsar, then, there should also be such an expanding cloud of dust and gas—one that was much smaller than the Crab Nebula in size, to be sure, since it would be so new, but one that would be much more active.

There is no sign of anything of the sort, and that must mean the supernova took place so long ago that the cloud it produced has long since dispersed to indetectability. That would make the Millisecond Pulsar quite old.

But now we're getting mixed signals. The ultrafast spin says "very young" and the absence of a nebula says "quite old." Which is it? How to decide?

One way is to determine the rate of slowing of the rotation speed. In the case of all the pulsars discovered prior to November 1982 the rule held that the faster the spin, the faster the rate of slowing.

The Millisecond Pulsar was therefore watched from day to

day and from week to week, and the rate of rotation carefully measured over and over again.

Astronomers found themselves utterly astonished. The Millisecond Pulsar was slowing at the rate of 1.26×10^{-19} seconds per second. This was a far smaller slowing effect than that of any other pulsar known, even though the rate of spin was far faster than that of any other pulsar known. The rate of slowing of the Crab Nebula pulsar is 3,000,000 times as great as that of the Millisecond Pulsar, even though the former rotates at less than 5 percent the speed of the latter.

Why is this? The general feeling is that the slowing effect arises because of the energetic emission of particles and radiation by a pulsar against the resistance of its own enormously intense magnetic field. If the Millisecond Pulsar slows down scarcely at all, it must have a very weak magnetic field and that should be the sign of an old pulsar. What's more, measurements seem to indicate that the surface temperature of the Millisecond Pulsar is less than 1,500,000 degrees K., which is very high by the standards of an ordinary star but quite low in comparison to all the other pulsars—again a sign of great age.

All the tests but one, then—the lack of a nebula, the low temperature, the weak magnetic field, the very slow rate of slowing of spin—seem to indicate an old pulsar. In fact, from its slowing rate astronomers guess that the Millisecond Pulsar may be 500 million years old (and perhaps older). Ordinary pulsars last only 10 to 100 million years before slowing and weakening to the point where the pulses can't be detected. The Millisecond Pulsar is already much older than what was thought to be the maximum pulsar lifetime and, considering its slow rate of energy loss, it has the potentiality of living on for billions of additional years.

But why is that? Most of all, why should such an old pulsar spin as though it were newborn?

The best guess so far is that the Millisecond Pulsar, having been formed long ago and having slowed and weakened to indetectability (many millions of years before there was anyone on Earth to detect it), was somehow revved up again comparatively recently.

Suppose, for instance, the pulsar was originally part of a binary system. (There are known cases of binary systems where one or both of the two stars is a pulsar, as I mentioned at the end of Chapter 4.

Some time after the pulsar had grown old and dim, the normal star that was its partner entered the red giant stage and expanded. The outer regions of the new red giant overflowed into the gravitational influence of the pulsar and formed an "accretion disc" of matter that was in orbit about the pulsar.

The weaker the magnetic field of the pulsar, the closer the accretion disc would be to the pulsar, and the faster the material in orbit would move under the gravitational whip of the tiny star.

The material in the accretion disc, at its inner edge, would spin about the pulsar faster than the aged, slow pulsar would be spinning about its own axis. The result would be that angular momentum would shift from the accretion disc to the pulsar. The pulsar would speed its spin and the accretion disc would slow down.

As the matter in the accretion disc slowed down, it would spiral inward toward the pulsar and speed up again, again transferring angular momentum to the pulsar. Eventually the material would spiral down into the pulsar, while new material would be entering the outer edge of the accretion disc from the companion star. Eventually a good part of the matter of the companion star would have bled onto the pulsar and the old pulsar would have increased its spin rate to the millisecond range. The companion would eventually be gone or would be too small in mass to maintain its nuclear fires, settling down as a black dwarf—a large planet, in fact.

The slow addition of the matter of the companion star to the pulsar would not restore its youth. The pulsar would still lack a nebula; it would still be cool and have a weak magnetic field; and because it had a weak magnetic field, it would still have a very low rate of slowing of spin. But it would have a very fast spin, just as though it were young.

If this suggestion is correct, and some astronomers argue strongly against it, it should not be a very uncommon scenario. Binary systems are extremely common—more common than single stars like our Sun. That means that most supernovas should be part of binary systems, and the resulting pulsars should, more often than not, have a normal star as companion. And if a binary system includes a pulsar, then every once in a while the normal star should evolve in such a way as to immolate itself on the pulsar and speed it up. For that reason, a close search of the heavens should uncover other old but speedy pulsars, perhaps even dozens of them.

There remains an interesting matter.

The Millisecond Pulsar has a rotation period that is the most delicately measured time interval we know. The rotation period has been measured to the nearest quadrillionth of a second (fifteen decimal places!) and with time we might be able to do a bit better than that.

Other pulsars are good clocks, too, but they are subject to periodic sudden small changes in the rotation rate (''glitches'') which may arise through internal changes in pulsar structure, or from the arrival of a sizable chunk of outside matter. This introduces an unpredictable inaccuracy in the ordinary pulsar clock. For some reason, there seem to be no glitches in the Millisecond Pulsar.

To be sure, the Millisecond Pulsar's rate is not constant. It

ASTRONOMY

is slowing perceptibly. Every 9¼ days its rate of spin becomes 1 quadrillionth of a second longer. This isn't much really, since it would take 2.5 billion years for its spin to become a billionth of a second longer if this slowing remained constant. Such a slowing rate could easily be allowed for.

What's the use of such a clock?

Well, to take one example, the Millisecond Pulsar can be used to time the passage of the Earth about the Sun. The irregularities in that passage—the small drifts ahead and the small lags behind the theoretical position if Earth and Sun were alone in the universe—could be measured more accurately than ever before.

These drifts would be due, in large part, to the perturbations induced on Earth by the other planets. These perturbations, in their turn, would depend on the mass of these planets and on their changing positions with time.

Knowing the positions of the planets through direct observation, and with better precision than ever thanks to the Millisecond Pulsar clock, we might well be able to calculate the mass of the various planets with a higher degree of accuracy than has been possible hitherto—especially that of the outermost planets, Uranus and Neptune.

And it is quite conceivable that applications even closer to home might arise, too.

PART III

CHEMISTRY

9

THE PROPERTIES OF CHAOS

Back in 1967 I wrote a book on photosynthesis, and it's just possible that some of you may interrupt me at this moment to ask me what the devil photosynthesis might be. If so, have faith! I will explain before this chapter is over.

I recognized the fact, at the time, that the five-syllable word was not one to inspire love and confidence at first sight, and it was my intention to give the book some dynamic title that would grab the reader's attention and get him to buy the book before quite realizing that it was full of moderately difficult biochemistry.

I didn't have the exact title in mind, so for a working title I let my imagination take a well-earned rest and used "Photosynthesis." By the time I had finished, I still did not have the exact title in mind so I thought I would let the publisher, Arthur Rosenthal of Basic Books, worry about it.

In 1968 the book was published and I received an advance copy and found, somewhat to my distress, that the title on the book jacket was *Photosynthesis*. In fact, believe it or not, that title was used *four times*.

I said tremulously, "Arthur, how do you expect to sell a book with the title *Photosynthesis—Photosynthesis—Photosynthesis—Photosynthesis?*"

And he said, "But haven't you noticed what else there is on the book jacket?"

"What?" said I, puzzled.

He pointed to the bottom right corner of the jacket where it said, clearly, *Isaac Asimov*.

As some of you may know, flattery always works with me, so I went off grinning, and, at that, the book did reasonably well. The publisher did not lose money on it, but I'll be frank with you—it was *not* a runaway bestseller.

So it has occurred to me to deal with some aspects of the subject once again in the lovably informal style I use in these chapters, and this time I will use a dramatic title, though I suppose that that alone will not make this book a runaway bestseller, either.

Let us start with the matter of eating. Animals, from the smallest worms to the largest whale, cannot live without food, and the food is, in essence, plants. All of us, from quadrillions of insects to billions of human beings, chomp away endlessly and remorselessly at the plant world, or at animals that have eaten plants; or at animals that have eaten animals that have eaten plants; or at—

Trace back the food chains of animals, and at their ends you will always find plants.

Yet the plant world does not diminish. Plants continue to grow as endlessly and as remorselessly as they are eaten but, as nearly as we can make out by simple nonscientific observation, they themselves do not eat. To be sure they require water, and sometimes they have to be helped along by care-

THE PROPERTIES OF CHAOS

fully larding the soil with something like animal excrement—
but we hesitate to consider that "eating."

In prescientific times it seemed to make sense to suppose
that plants were an order of objects that were totally different
from animals. Of course plants grew as animals did, and were
produced from seeds as some animals were produced from
eggs, but these seemed comparatively superficial likenesses.

Animals moved independently, breathed, and ate. Plants
did none of these things, any more than rocks did. Independent motion, in particular, seemed an essential property of
life, so that whereas all animals were self-evidently alive,
plants (like rocks) were not.

This would seem to be the view of the Bible. When the dry
land appeared on the third day of the Genesis account of
creation, God is described as saying, "Let the earth bring
forth grass, the herb yielding seed, and the fruit tree yielding
fruit." (Genesis I:11).

No mention is made of life being characteristic of the plant
world.

It is not till the fifth day that life is mentioned. Then God
says, "Let the waters bring forth abundantly the moving
creature that hath life . . . And God created great whales,
and every living creature that moveth . . ." (Genesis I:20–21).

Animals are characterized as moving and alive, the two
apparently implying each other. Plants are neither.

God says: ". . . to every beast of the earth, and to every
fowl of the air, and to every thing that creepeth upon the
earth, wherein there is life, I have given every green herb for
meat . . ." (Genesis I:30). In other words, the moving animals are alive and the nonmoving plants are merely food
supplied them by the grace of God.

Herbivorousness is clearly considered the ideal. Carnivo-
rousness is not mentioned in the Bible till after the Flood,
when God tells Noah and his sons, "Every moving thing that

liveth shall be meat for you; even as the green herb have I given you all things.'' (Genesis 9:3).

In general, Western thought followed the words of the Bible (as it could not help but do, since the Bible was considered the inspired word of God). The nonliving, nonnourishing ground was somehow converted into nonliving but nourishing plants which could serve as food for living animals. The seed, when sown, served as a triggering agent for the ground-to-plant conversion.

The first person to test this theory of plant growth was a Flemish physician, Jan Baptista van Helmont (1580–1644). He planted a young willow tree weighing five pounds in a pot containing 200 pounds of earth. For five years he let the willow tree grow, watering it regularly and covering the earth carefully between waterings so that no extraneous matter could fall into it and confuse the results.

After five years, he withdrew the now much larger willow tree from the pot and carefully knocked off all the earth adhering to the roots. The willow tree weighed 169 pounds, having thus gained 164 pounds. The earth had lost, at most, one eighth of one pound.

This was the first quantitative biochemical experiment we know of, and was of crucial importance for that, if for nothing else. In addition it showed conclusively that earth did not convert itself, to any but the tiniest degree, into plant tissue.

Helmont reasoned that since the only other material that entered the system was water, the willow tree and presumably plants generally were formed out of water.

The reasoning seemed iron-bound, especially since it had been well known from earliest times that plants simply would not grow if they were deprived of water.

And yet the reasoning was wrong, because water was *not*

the only material other than earth that touched the willow tree. The tree was also touched by air, and Helmont would instantly have acknowledged the fact if it had been pointed out to him. It was just that air, being invisible, impalpable, and apparently immaterial, was always easy to ignore. Helmont had other reasons to do so, too.

In Helmont's time, air and related substances were beginning to be studied scientifically for the first time. Indeed, it was Helmont himself who initiated the process.

Thus, earlier chemical experimenters had noticed, and reported, vapors that formed in their mixtures and came bubbling upward, but had dismissed them as varieties of air.

Helmont was the first to study these "airs" and to note that they sometimes had properties quite distinct from that of ordinary air. Some of the vapors were, for instance, inflammable, which ordinary air certainly was not. Helmont noted that when such inflammable vapors burned, droplets of water sometimes formed.

Nowadays, of course; we know that when hydrogen burns it forms water, and we can be pretty sure that that was what Helmont observed. Helmont, without the advantage of our hindsight brilliance, came to the rather simpler conclusion that such an inflammable vapor (and therefore all vapors, even including ordinary air itself) was a form of water. Therefore he naturally dismissed air as the source of the willow tree's substance. It was water that was the source, whether in liquid or vaporous form.

Helmont noted that liquid water had a definite volume, whereas vapors did not. Vapors expanded to fill spaces, interpenetrating everything. They seemed to lack order, to be substances that were in total disorder.

The Greeks believed that the universe began as a kind of substance that was in total disorder. The Greek term for this original, disorderly substance was "Chaos." Helmont called his vapors by this term, using his own Flemish pronunciation, which, when spelled phonetically, produced the word "gas." To this day, we call air a gas, and we apply the word to any vapor or airlike substance.

Helmont studied the properties of chaos—that is, the properties of gases. He produced a gas from burning wood that was not inflammable and that tended to dissolve in water (something that Helmont would naturally interpret as being converted into water). He called it "gas sylvestre" ("gas from wood") and it is the gas we know today as carbon dioxide. It is a pity Helmont had no way of knowing the significance of the discovery in connection with his investigation of the problem of plant growth.

The study of gases took another leap forward when an English botanist, Stephen Hales (1677–1761), learned how to collect them with reasonable efficiency.

Instead of simply allowing them to escape into the air and thus being forced to study them on the fly, so to speak, he produced his gases in a reaction vessel with a long neck that curved downward and upward again. This long neck could be inserted into a trough of water, and the opening of the neck could be covered by an inverted beaker also full of water.

When a particular gas formed as a result of the chemical changes taking place in the reaction vessel, it bubbled to the surface of the reacting materials, filled the air space above, expanded through the long, curved neck and into the inverted beaker. The gas collected in the beaker, stayed put, and the properties of one particular chaos could be studied at leisure.

Hales prepared and studied, in this fashion, a number of gases, including those we now call hydrogen, sulfur dioxide, methane, carbon monoxide, and carbon dioxide. He didn't get enough out of it, however, for he persisted in thinking of them all as varieties of ordinary air.

It was impossible to work with these gases, however, without eventually coming to the conclusion that air was not a simple substance but a mixture of different gases.

A Scottish chemist, Joseph Black (1728–99), was interested in carbon dioxide and found, in 1756, that if it was brought into contact with the common solid substance called lime (calcium oxide), it was converted into limestone (calcium carbonate).

He then took note of a crucial fact. He didn't have to use laboriously prepared carbon dioxide for the purpose. He merely had to leave lime in contact with ordinary air. The limestone would form spontaneously, although much more slowly than if he used carbon dioxide. Black's conclusion was that air contained carbon dioxide in small quantities, and in that conclusion he was completely correct.

In 1772 another Scottish chemist, Daniel Rutherford (1749–1819), a student of Black's, allowed candles to burn in a closed container of air. Eventually the candle would no longer burn, and what's more neither would other things burn in that air. Nor would a mouse live.

By this time it was known that a burning candle produced carbon dioxide, so it was an easy conclusion that all the normal air that would allow burning had been replaced by carbon dioxide, which was known not to allow burning.

On the other hand, it was also known that carbon dioxide would be absorbed by certain chemicals (such as lime). The air in which the candle had burned was passed through these chemicals and, indeed, carbon dioxide was removed. Nevertheless most of the air remained untouched, and what was left, though not carbon dioxide, would not support combustion. What Rutherford had isolated was the gas we now call nitrogen.

An English chemist, Joseph Priestley (1733–1804), also studied gases. He studied the gas that was produced by

CHEMISTRY

fermenting grain (he lived next door to a brewery) and found that it was carbon dioxide. He studied its properties, in particular the manner in which it dissolved in water, and discovered that a solution of carbon dioxide produced what he considered (but I do not) a pleasant, tart drink.

(When I was young, such carbonated water was called seltzer and could be bought for a penny a glass. Nowadays it is called Perrier and can be bought, I believe, for a dollar a glass. I refused, in my youth, to invest a penny in the sour drink, and I doubly refuse to invest a dollar today.)

Priestley was the first to lead gases through mercury rather than water and was thus able to collect some gases which would have dissolved instantly in water, using Hales's method. In this way Priestley isolated and studied such gases as hydrogen chloride and ammonia.

His most important discovery came in 1774. When mercury is strongly heated in air, a brick-red powder forms on its surface. This is the result of mercury combining (with some difficulty) with a portion of the air. If the brick-red powder is collected and heated again, the mercury-air combination is broken up and the air component is liberated as a gas.

Priestley discovered that this air component supported combustion with great ease. A smoldering splint burst into active flame if placed in a beaker containing this gas. Mice penned up in a container of it behaved in an unusually frisky manner, and when Priestley breathed some of it, it made him feel "light and easy." It is the gas we now call "oxygen."

It was the French chemist Antoine Laurent Lavoisier (1743–94), by common consent the greatest chemist who ever lived, who made sense of this. His careful experiments showed him, by 1775, that air consisted of a mixture of two gases, nitrogen and oxygen, in an approximately 4-to-1 ratio by volume. (We know now that there are a number of minor

142

constituents in dry air, making up about 1 percent of the whole, and that 0.03 percent carbon dioxide is included.)

Lavoisier showed that combustion is the result of the chemical combination of substances with the oxygen of the air. For instance, the burning of coal, which is almost pure carbon, is its combination with oxygen to form carbon dioxide. When hydrogen burns, it combines with oxygen to form water, which thus consists of a chemical combination of these two gases.

Lavoisier correctly suggested that the food we eat and the air we breathe combine with each other, so that respiration is a form of slow combustion. This means that human beings inhale air that is comparatively rich in oxygen, but exhale air that is comparatively depleted in that gas and enriched in carbon dioxide. Careful chemical analysis of exhaled air showed this to be true.

There was now a satisfactory explanation for the fact that a candle burning in a closed container of air eventually stopped burning, that a mouse living in such a chamber eventually died, and that the remaining air in either chamber would not support the burning of any other candle or the breathing of any other mouse.

What happened was that either burning or respiration gradually consumed the oxygen content of air and replaced it with carbon dioxide, leaving the nitrogen unchanged. Air made up of a mixture of nitrogen and carbon dioxide could support neither combustion nor respiration.

That brought up an interesting problem. Every animal living is constantly respiring, constantly inhaling air that is 21 percent oxygen, and constantly exhaling air that is only 16 percent oxygen. Surely there would come a time when the oxygen content of Earth's atmosphere as a whole would be depleted to the point where life would become impossible.

This should have happened in less time than is included in

the known history of civilization, so we can only conclude that something is replacing the oxygen as fast as it is used up. But what is the something?

The first hint of an answer to the problem came from Priestley even before he discovered oxygen.

Priestley had penned up a mouse in a closed container of air, and eventually the mouse died. The air as it then existed would not support the life of any other animal, and Priestley wondered whether it would also kill plants. If it did, that would show that plants, too, were a form of life, which would be an interesting, though unbiblical, conclusion. (The unbiblicality of it would not have bothered Priestley, who was a Unitarian and therefore a religious radical—and a social radical as well, by the way.)

In 1771 Priestley put a sprig of mint into a glass of water, and put it into a container of air in which a mouse had lived and died. The plant did not die. It grew there for months and seemed to flourish. What was more, at the end of that time a mouse could be placed in the enclosed air and it would live for a considerable while, and a candle placed in it would continue to burn for some time.

In short, the plant seemed to have revitalized the air which the animal had depleted.

In modern terms we would say that whereas animals consume oxygen, plants produce it. The combination of the two processes left the overall percentage of oxygen in the atmosphere unchanged.

Plants thus perform the double service of supplying animal life with an endless supply of oxygen as well as food, so that although animals (including you and me) breathe and eat constantly, there is always more oxygen and food in existence to breathe and to eat.

Once Lavoisier explained combustion and put chemistry on

its modern foundations, the matter of plant activity roused particular interest.

A Dutch botanist, Jan Ingenhousz (1730–99), heard of Priestley's experiment and decided to go more deeply into the matter. In 1779 he performed many experiments designed to study the manner in which plants revitalized used-up air, and discovered that plants produced their oxygen only in the presence of light. They did so by day, but not by night.

A Swiss botanist, Jean Senebier (1742–1809), in 1782 confirmed Ingenhousz's findings and went farther. He showed that something else was necessary for the production of oxygen by plants: they had to be exposed to carbon dioxide, as well.

The time was now ripe to repeat Helmont's experiment of a century and a half before, in the light of new knowledge. This was done by another Swiss botanist, Nicolas Théodore de Saussure (1767–1845). He allowed plants to grow in a closed container with an atmosphere containing carbon dioxide and carefully measured how much carbon dioxide was used up by a plant and how much weight of tissue it gained. The gain in tissue weight was considerably greater than the weight of carbon dioxide used up, and de Saussure showed quite convincingly that the only possible source of the remaining weight was water.—Helmont had been partly right.

By now enough was known to make it clear that plants were as alive as animals were and to get an idea of how the two great branches of life balanced each other.

Food, whether of plant or animal tissue, is rich in carbon and hydrogen atoms, C and H. (The atomic theory was established in 1803 and was adopted by chemists rather rapidly.) When food was combined with oxygen, it formed carbon dioxide (CO^2) and water (H^2O).

The combining of substances containing carbon and hydrogen atoms with oxygen atoms generally liberates energy. The chemical energy of the carbon-hydrogen substances is converted in the body to kinetic energy, as when muscles contract, or into electrical energy, as when nerves conduct impulses, and so on. We might therefore write:

food + oxygen → carbon dioxide + water + kinetic (etc.) energy

With plants it is just the other way around:

Light + carbon dioxide + water → food + oxygen

What it amounts to is that plants and animals, working together, keep food and oxygen on one side and carbon dioxide and water on the other in balance, so that, on the whole, all four remain constant in amount, neither increasing nor decreasing.

The one irreversible change is the conversion of light energy into kinetic, etc., energy. That has been going on for as long as life has existed and can continue on Earth as long as the Sun continues to radiate light in approximately the present fashion. This was first recognized and stated in 1845 by the German physicist Julius Robert von Mayer (1814–78).

How did this two-way balance come to evolve?—We can speculate on the matter.

Originally it was the ultraviolet light of the Sun that probably supplied the energy for building up relatively large molecules out of the small ones in the primordial sea's lifeless waters. (The conversion of small molecules to large ones usually involves an input of energy; the reverse usually involves an output of energy.)

Eventually, when molecules large and complex enough to possess the properties of life were formed, these could use (as food) molecules of intermediate molecules (complex enough to yield energy on breakdown, but not complex enough to be alive and capable of fighting back).

The Sun's energy, working on a hit-and-miss basis, formed

only so much in the way of intermediate molecules, however, and these could support only so much life.

It paid living systems, therefore, to form membranes about themselves (becoming cells) which could allow small molecules to pass inward. If the living systems possessed devices that would make use of solar energy for molecular buildup, those small molecules would be built up to large ones before they had a chance to get out again—and the large ones, once formed, could not get out either.

In this way, these cells (the prototypes of plants) would live in a microenvironment rich in food and would flourish to a far greater extent than precellular life-forms that lacked the ability to direct food manufacture through the use of solar energy.

On the other hand, cells lacking the ability to use solar energy to make food could still flourish if they developed means to filch the food content of cells that could. These filchers were the prototypes of animals.

And are these filchers parasites and nothing more?

Perhaps not. If plants existed alone they would concentrate all available small molecules into their own tissues, and growth and development thereafter would be slow. Animals serve to break down a reasonable proportion of the complex contents of plant cells and allow continued plant growth, development, and evolution at a greater rate than would otherwise be possible.

Food molecules are far larger and more complex than the molecules of carbon dioxide and water. The two latter have molecules made up of three atoms each, whereas the characteristic molecules of food are made up of anywhere from a dozen to a million atoms.

Putting together large molecules from many small ones is called "synthesis" by chemists, from Greek words meaning

CHEMISTRY

"to put together." Whereas animals characteristically break down food molecules by combining them with oxygen to form carbon dioxide and water, plants characteristically synthesize food molecules out of carbon dioxide and water.

Plants do it by making use of the energy in light. Therefore that particular kind of synthesis is called "photosynthesis," the prefix, "photo-" being from the Greek word for "light." —Didn't I tell you I'd explain the word?

And there are a few more things I can say about it, too—next chapter.

10

GREEN, GREEN, GREEN IS THE COLOR . . .

When I was buying the electric typewriter on which the first draft of this chapter is now being written (final copy will be on my word processor), the salesman placed his final question to me.

"And what color would you like?" he asked, and presented me with a page on which the various colors were illustrated in most lifelike fashion.

For me this was a troublesome question, for I am not visually oriented and I generally don't care what color things might be. Looking over the samples thoughtfully, it struck me that I had had a typewriter of every indicated color but green. I therefore asked for green and eventually the typewriter arrived.

Whereupon Janet (my dear wife) registered amazement. "Why did you pick green?" she asked.

I explained.

She said, "But your carpet is blue. Or haven't you noticed?"

I looked at the carpet, which I've only had for seven years

and, holy smokes, she was right. I said, "Does it make a difference?"

"Most people," she said, "would think that green and blue clash."

I thought about it and said, "The grass is green and the sky is blue and people are always talking about the beauties of nature."

For once I had her. She laughed and never said another word about my green typewriter.

I, for my part, however, intend to spend some time talking about green.

In the preceding chapter I explained that animals combine the complex molecules of food with the oxygen of the air and in the process break down those complex molecules to the relatively simple ones of carbon dioxide and water. The energy liberated by this means is utilized by the animal body in all the energy-consuming processes characteristic of life— muscle contraction, nerve impulse, gland secretion, kidney action and so on.

Plants, on the other hand, make use of the energy of sunlight to reverse the above process (photosynthesis), combining carbon dioxide and water to form the complex molecules of the type found in food, and liberating oxygen in the process.

Plants and animals, taken together, mediate a cyclic chemical process that keeps complex molecules, oxygen, water, and carbon dioxide all in a steady state. The one permanent change is that of the conversion of solar energy into chemical energy.

The question is: What makes plants and animals so different? What is there about plants that makes it possible for them to photosynthesize, making use of the energy of sunlight to do so; and what is there about animals that makes it impossible for them to do the same?

Before we delve into the depths of cells and molecules in search of something very subtle and delicate, we might as well step back and see if there is, by some chance, something very noticeable that would answer our question.

It might seem we might not have much chance of finding something immediately on the surface, since Mother Nature tends to keep her little tricks under her hat, but in this case a very obvious point shows up at once.

One thing that springs to the eye is that all plants, or at least important parts of all plants, are green! What's more, while animals may sport a variety of colors, green is conspicuous by its absence.

Neither statement is completely universal (and I had better say that before some reader does). There are living things that resemble plants in very many ways—growing from the ground, possessing cellulose, and demonstrating various other physical and chemical properties associated with plants—that are nevertheless not green. The most familiar examples are mushrooms and toadstools, and such nongreen plants are lumped together as "fungi," from a Latin word for mushrooms.

In the same way there are parrots that, although undoubtedly animals, have plumage of a striking green. (There is, however, no chemical similarity whatsoever between the green of parrot feathers and the green of grass.)

Such exceptions are trivial and do not detract from the importance of the generalization that plants are green and animals are not green.

Perhaps, however, this is coincidence; and perhaps the two contrasts—green vs. not-green, and photosynthesis vs. non-photosynthesis—have nothing to do with each other.

Not so! Where plants are in part green and in part not-green, it is invariably in the green portion that photosynthesis takes place. Thus in a tree it is in the green leaves that we find photosynthesis, and not in the brown bark or in the multihued flowers. And in fungi, which are plants with no green parts, there is also no photosynthesis. Fungi, like animals, can grow only if complex molecules in one way or another are already available to them.

For that reason we often speak of photosynthesis as taking place not in plants but in *green* plants, thus making certain we do not overgeneralize.

Why should color have anything to do with photosynthesis?— Remember that the process requires the use of solar energy.

If sunlight passed right through a plant, none of it could be used to supply the necessary energy. The same would be true if the sunlight were all reflected. In the first case the plant would be transparent, and in the second case it would be white, and in neither case would it photosynthesize.

In order for photosynthesis to take place, sunlight must be stopped and absorbed by the plant. If all the sunlight were absorbed, the plant would be black, but such total absorption is not necessary.

Sunlight is a mixture of an enormous number of different wavelengths of light, with each wavelength made up of quanta containing a specific energy content. (The longer the wavelength, the smaller the energy content of the quanta.)

For a particular chemical change to take place, a particular amount of energy must be supplied, and those quanta work best that produce just the right amount. In the case of photosynthesis, it is the red light that works best, and this is a good thing. Red light has the longest wavelengths of visible light and it can penetrate mist and clouds somewhat better than other forms of visible light, and is less scattered when the sun is low in the sky. Plants do better, therefore, to depend on red light rather than any other form of visible light.

In that case, why bother evolving a photosynthetic system that absorbs anything more than the red light? To absorb shorter wavelengths would serve no purpose, would require the evolution of special compounds with the necessary capability, and would unnecessarily raise the temperature of the plants.

Plants have a photosynthetic system, therefore, that tends to absorb the red portion of the sunlight and to reflect the rest. Reflected sunlight minus the red portion that is absorbed is green in color, so plants that photosynthesize are naturally green, and it is to be expected that plants that are green might well be capable of photosynthesis. The two, greenness and photosynthesis, have a logical connection, and the fact that one is accompanied by the other is no coincidence.

We have to go beyond mere greenness, however.

If a piece of plant tissue is green, this is only because some specific chemical within that tissue absorbs the red light, reflecting the rest, and that specific chemical is therefore itself green.

Two French chemists, Pierre Joseph Pelletier (1788–1842) and Joseph Bienaimé Caventou (1795–1877) were particularly interested in isolating chemicals of biological importance from plants. Among the chemicals they were the first to isolate, between 1818 and 1821, were alkaloids such as strychnine, quinine, and caffeine.

Even before that, in 1817, they had extracted material that contained the green coloring matter of plants and were the first to give that substance its name. They called it "chlorophyll" which comes from Greek words meaning "green leaf."

This advance is important, but it is only a beginning. Pelletier and Caventou might look at a green solution in a test tube and give it a name, but what *is* chlorophyll?

In 1817 the atomic theory was only a decade or so old and

chemists had no way of pinning down the arrangement of atoms within a complicated molecule. It was not till 1906 that the first major attack on the atomic structure of chlorophyll was made, and that was by the German chemist Richard Willstätter (1872–1942).

He was the first to prepare chlorophyll in reasonably pure form and discovered that it was not one but two very closely related chemicals which he called "chlorophyll-a" and "chlorophyll-b," the two differing slightly in their patterns of light absorption. The former was the more common, making up about three forths of the mixture.

Having the chemicals pure, he was able to study the different elements present with considerable confidence that those elements would prove to be actually part of the chlorophyll molecules, and not part of any impurities that might also be present. Chlorophyll contains atoms of carbon, hydrogen, oxygen and nitrogen, but that was no surprise. By Willstätter's time it was known that virtually every complex molecule in living organisms (so-called organic molecules) contained carbon, hydrogen, and oxygen atoms, and that a good percentage of them contained nitrogen atoms as well.

Willstätter found, however, that chlorophyll contained magnesium atoms too. It was the first organic molecule discovered to contain that element.

Nowadays we know that each molecule of chlorophyll-a contains 137 atoms, while each molecule of chlorophyll-b contains 136 atoms.

A molecule of chlorophyll-a is, at present, known to have 72 hydrogen atoms, 55 carbon atoms, 5 oxygen atoms, 4 nitrogen atoms, and 1 magnesium atom. A molecule of chlorophyll-b has two fewer hydrogen atoms and one more oxygen atom.

If one knows the total number of atoms in a molecule, and how many of each variety are present, that still amounts to very little. What counts is the arrangement of these atoms, and 136 or 137 atoms of five different kinds can be arranged in an astronomical number of different ways.

One way of getting some hint of the arrangement is to break up the complex molecules, somehow, into simpler fragments which one can then study. A particular fragment might contain no more than a dozen or so atoms of three different kinds, and might fit together reasonably well in only two or three different ways. It might even be that chemical experience would lead one to suppose that the chance of one particular arrangement of those few atoms was much more likely than any other.

Then, to settle the matter, chemists could actually synthesize various molecules containing the requisite number of different kinds of atoms in each of the different arrangements that are at all likely, comparing each one with the fragment obtained from the chlorophyll molecule. When an identity of properties shows up, you know that your fragment is equivalent to the synthesized compound that it matches.

In this way, Willstätter discovered that among the fragments of the chlorophyll molecules were to be found small molecules containing four carbon atoms and a nitrogen atom, these five atoms being arranged in a ring. The simplest such ring has one hydrogen atom attached to each of the five atoms of the ring. This compound was named "pyrrole" by the German chemist Friedlieb Ferdinand Runge (1795–1867), who first isolated it in 1834. The name is from a Greek word for a fiery red, since when pyrrole is treated with certain acids a bright red substance is formed.

It seemed logical, therefore, to suppose that chlorophyll consisted of pyrrole rings arranged in some fashion that produced a still more complicated pattern. In 1912 a chemist named William Küster proposed that four pyrrole rings could form a larger ring, each pair of pyrroles being connected by a bridge consisting of a single carbon atom.

A compound made up of such a ring of pyrrole rings is called a "porphyrin," a term first used by the German biochemist Felix Hoppe-Seyler (1825–95) in the 1860s. Porphyrin comes from a Greek word for purple, since many porphyrins are purple in color.

By the time Willstätter's work was done, then, it seemed fairly certain that chlorophyll possessed a molecule that had a porphyrin ring at its core, but there were still plenty of details that required elucidation.

The porphyrin ring-of-rings possesses many symmetries in the arrangement of the atoms, and these symmetries contribute to the stability of the molecule. (The American chemist Linus Pauling (1901–   ) demonstrated this sort of thing in his revolutionary application of quantum mechanics to molecular structure fifty years ago.) Consequently the porphyrin structure, with its ring skeleton made up of 20 carbon atoms and 4 nitrogen atoms, is commonly found in life and is included in various essential compounds in both plants and animals, and not in chlorophyll alone.

Thus, in many animals (including human beings) there is a purplish porphyrin, "heme." This heme, when attached to an appropriate protein, forms hemoglobin, the red substance that absorbs oxygen at the lungs, or gills, and delivers it to the tissue cells. In combination with other proteins, heme forms enzymes involved in the handling of oxygen by the cells and these are found, universally, in all oxygen-using cells, plant and animal.

It is a sample of the economy of nature that the same stable ring of rings can, with slight modification, produce the green chlorophyll so essential to plants and the heme so useful to animals. (Whereas, in chlorophyll, the green color is of the essence, in heme the purple color is a mere side issue and plays no part in its functioning.)

But how is the porphyrin ring modified to form this compound or that?

The four pyrrole rings are arranged with the nitrogen atoms pointing toward the center. The two carbon atoms adjacent to the nitrogen atom in each pyrrole ring are involved in the large ring formation (those carbon atoms are what the pyrrole rings ''hold hands'' with).

That leaves the two carbon atoms at the far end of the pyrrole ring free. These eight carbon atoms (two in each of the pyrrole rings) can each be attached to a side chain consisting of one or more carbon atoms to which, in turn, still other atoms may be attached. What particular side chains are involved, then, and where on the porphyrin ring does each side chain go?

The problem was tackled by the German chemist Hans Fischer (1881–1945) in the 1920s. He worked with heme, and after knocking off the side chains he studied and analyzed the resulting mixture. He demonstrated that each heme molecule has four side chains consisting of 1 carbon atom and 3 hydrogen atoms (a ''methyl group''); two more consisted of 2 carbon atoms and 3 hydrogen atoms (a ''vinyl group''), and two side chains consisting of 3 carbon atoms, 5 hydrogen atoms, and 2 oxygen atoms (a ''propionic acid group'').

These eight groups of three different varieties can be arranged about the porphyrin-ring skeleton in fifteen different ways. Which way is correct?

Fischer had developed methods for synthesizing porphyrin molecules complete with side chains, and he therefore adopted a strategy of mass assault. He told each of fifteen graduate students to synthesize a different porphyrin molecule with the side chains arranged in a particular fashion, so that among them all fifteen would be produced. By 1929 he was able to show that a particular one of the fifteen was correct. The side-chain arrangement, as one goes around the porphyrin ring, turned out to be methyl, vinyl, methyl, vinyl, methyl, propionic acid, propionic acid, methyl.

Fischer then went on to tackle chlorophyll. Obviously there would have to be differences, the major one being that heme had an iron atom at the center of the porphyrin ring, while chlorophyll had a magnesium atom. However, if the iron atom was knocked out of the first and the magnesium atom out of the second, what was left in the two cases were not identical. There were other differences as well.

To begin with, the four methyl groups are in the same place in the chlorophyll-porphyrin as in the heme-porphyrin. So are the two vinyl groups, except that in the second one, two additional hydrogen atoms are added to form an "ethyl group." The propionic acid groups are in the same place as in heme but are considerably modified. One of the propionic acid groups curls around to combine with the adjacent pyrrole ring to form a fifth ring, and an additional carbon atom is added to it. In the case of the other, it dangles free but a long twenty-carbon chain (the "phytyl group") is attached to it.

At least, that's chlorophyll-a. In chlorophyll-b, one of the methyl side chains is converted into an "aldehyde group," which is made up of one atom each of carbon, hydrogen, and oxygen.

This description of the chlorophyll structure was deduced from Fischer's analyses of fragments, but the final confirmation could not come until a chlorophyll structure was built up in the laboratory, one that matched the suggested structure. If the synthetic molecule then proved to be identical in all properties to the natural one, the structure would then be proved beyond a reasonable doubt.

Chlorophyll has a more complicated structure than heme does, however, and its synthesis eluded Fischer. Nor was it successfully achieved till 1960, when the American chemist Robert Burns Woodward (1917–79) managed the job—and the structure was confirmed.

Once we have chlorophyll, and can even synthesize it, does the possibility arise that we can short-circuit the plant world? Perhaps we can isolate chlorophyll and set it to work in large chemical plants. Shining light upon it and supplying it with favorable conditions, might we get it to convert carbon dioxide and water into food substances with high efficiency and without the necessity of its wasting effort in supplying the needs of plant structure and functioning?

No! Put chlorophyll in a test tube and expose it to light and it does *not* photosynthesize. Even if you extract other compounds also present in plant cells and add them to the chlorophyll there will be no photosynthesis taking place. Apparently, within plant cells, chlorophyll is part of an intricate and well-organized system that acts as a smoothly working whole to carry through a photosynthetic process including many steps. Chlorophyll makes the key step possible, and without it nothing can happen, but the key step by itself is also not enough.

An organism is made up of cells, but each cell is not an unorganized drop of protoplasm. Rather, within each cell are still tinier structures called organelles, each one of which is itself highly organized. As examples, one important type of organelle present in virtually all cells are chromosomes, which contain the genetic machinery that makes reproduction possible. Another are the mitochondria, which are the powerhouses of the cell and contain a complex enzyme system so organized as to make it possible to combine food and oxygen in such a way as to produce energy in a controlled and useful fashion.

Within plant cells chlorophyll, it turns out, is also confined to certain organelles. This was first demonstrated in 1865 by the German plant physiologist Julius von Sachs (1832–97). These organelles were eventually named chloroplasts.

Chloroplasts are large organelles—two to three times as long and as thick as mitochondria, for instance—and it is not surprising that the chloroplast structure is correspondingly the more complex of the two.

The interior of the chloroplast is made up of many thin membranes stretched across the width of the organelle. These are the lamellae. In most types of chloroplasts, these lamellae thicken and darken in places to form condensations called grana. The chlorophyll molecules are to be found within the grana.

If the grana are studied under the electron microscope, they in turn seem to be made up of tiny units, just barely visible, that look like the neatly laid tiles of a bathroom floor. Each of these objects may be a photosynthesizing unit containing 250 to 300 chlorophyll molecules.

The chloroplasts are more difficult to handle than mitochondria are, for with increased structural complexity, it would seem, comes added fragility. When cells are broken up, for instance, mitochondria can be isolated, intact, with relative ease, and can even be made to continue to perform their function.

Not so with chloroplasts. Even gentle methods of extracting them from fragmented cells will succeed in destroying them. Even when they look intact, they are not, for they will not photosynthesize.

It was not till 1954 that chloroplasts sufficiently intact to allow them to carry through the complete photosynthetic reaction were obtained, by the Polish-American plant physiologist Daniel Israel Arnon (1910–   ), working with disrupted spinach leaf cells.

*   *   *

Is that the answer, then? Do we isolate chloroplasts rather than chlorophyll, and set them working in the laboratory under optimal conditions, and churning out starch, fat, and protein for us?

Theoretically, yes, but practically, no. In the first place, we would have to depend on the plant world for the supply of chloroplasts. In the second place, chloroplasts are so fragile that we would forever have to be renewing our supply. It would be enormously cheaper and more efficient in the long run to continue to use the chloroplasts where they can easily preserve and reproduce themselves—inside the intact and living plant cell.

But why try to duplicate photosynthesis on the plant's terms? Might we not find a substitute?

The key step in photosynthesis is the splitting of the water molecule to hydrogen and oxygen. Chemists can do that easily, but only at the cost of a large energy input. They can do it by heating the water molecules strongly enough to vibrate them to pieces, or by passing an electric current through a dilute solution of sulfuric acid so that electric charges can pull the molecules apart. Both the heat and electricity, however, represent a massive input of energy. The hydrogen we can isolate in this way would, when recombined with oxygen, liberate considerable energy that we can then utilize—but the energy liberated would be not nearly as great as the energy we expended to break up the water molecule and obtain the hydrogen in the first place.

Suppose, though, we could split the water molecule by making use of sunlight, as plants do. To be sure, the energy of sunlight would be greater than the energy we would then obtain by combining the liberated hydrogen with oxygen, but we would not have to invest anything to produce the sunlight. The sunlight is always there and would just go to waste if we did not use it.

Plants do this through their chloroplasts, but can *we* do it through some simpler system—stable, efficient, and working tirelessly according to our direction?

The hydrogen and oxygen we would form from water could be recombined to yield energy that would be more concentrated and useful than the original sunlight would be. In the process, water molecules would be reformed. No water, hydrogen, or oxygen would ever be consumed, and the only permanent change would be the conversion of dilute sunlight into concentrated chemical energy. The process could continue as long as the sun shone in its present manner.

What's more, once the hydrogen was formed, we might work out methods for combining it with carbon dioxide to form food. In this way, we could look forward to a future in which human beings, at need, could do without the plant world altogether. We would get food and fuel at the expense of sunlight.

Naturally I am not advocating the elimination of the plant world, but there might be times when we would temporarily have to do without it—on long voyages through space in ships not large enough for an ordinary ecological balance, for instance. It would be useful, then, if we could set up an artificial system that would turn the trick.

And chemists are on the track. The American biochemist Melvin Calvin (1911–   ) who, in 1961, obtained a Nobel Prize for his work in deciphering the details of the photosynthetic reaction, is using synthetic metal-containing compounds designed to mimic the activity of chlorophyll. Others are also in the field.

So far, no one has quite created the equivalent of an artificial plant cell, but there is no reason success should not come about eventually and make it possible for human beings to supplement their food and fuel supply in this way—and even, at need, to function for lengthy periods in a situation in which they themselves (plus their internal parasites) are the only living organisms.

PART IV
BIOLOGY

11

MORE THINKING ABOUT THINKING

In my book *The Planet That Wasn't* (Doubleday, 1976) there is an essay of mine entitled "Thinking About Thinking." In it I expressed my dissatisfaction with intelligence tests and gave my reasons therefor. I presented arguments for supposing that the word "intelligence" stood for a subtle concept that could not be measured by a single figure such as that represented by an "intelligence quotient" (IQ).

I was very pleased with the article, all the more so since it was attacked by a psychologist for whose work I have little respect (see "Alas, All Human" in *The Sun Shines Bright*, Doubleday, 1981).

Nor did I think I would ever have to add to it. In fact, I rather suspected I had emptied myself of all possible thought I might have on the subject of intelligence.

And then, not long ago as I write this, I found myself sitting at a dinner table with Marvin Minsky of M.I.T. on my right hand and Heinz Pagels of Rockefeller University on my left.

Pagels was conducting a three-day conference on computers and earlier that day had moderated a panel discussion entitled "Has artificial intelligence research illuminated human thinking?"

I did not attend the panel (various deadlines precluded that) but my dear wife, Janet, did, and from her account it would seem that Minsky, one of the panelists, and John Searle of the University of California engaged in a dispute on the nature of artificial intelligence. Minsky, the leading proponent of that field of research, opposed Searle's view that consciousness was a purely biological phenomenon and that no machine could ever have consciousness or intelligence.

At the dinner Minsky continued to maintain his view that artificial intelligence was *not* a contradiction in terms, while Pagels was supporting the legitimacy of Searle's view. Since I was sitting between, the polite but intense argument was being conducted over my head both literally *and* figuratively.

I listened to the arguments with increasing anxiety for I had carelessly agreed, months before, to give an after-dinner talk that night. It now seemed to me that the Minsky-Searle debate was the only topic on the collective mind of the high-powered dinner attendees and that it would be absolutely necessary for me to talk on that subject if I were to have any chance of holding their attention.

It meant that I had to go back to thinking about thinking, and that I had less than half an hour in which to do it. I managed, of course, or I wouldn't be telling you this. In fact, I was told that during the rest of the conference I was occasionally quoted with approval.

I can't give you my talk word for word, since I spoke off the cuff as I always do, but here is a reasonable facsimile.

Suppose we start with the easy assumption that *Homo sapiens* is the most intelligent species on Earth, either living now or in the past. It should not be surprising, therefore, that the human brain is so large. We tend, with considerable reason, to associate the brain with intelligence and vice versa.

The brain of the adult human male has a mass of about 1.4 kilograms, on the average, and is far larger than any nonmammalian brain, past or present. This is not surprising, considering that mammals as a class have larger brains and are more intelligent than any other kinds of living organisms.

Among the mammals themselves it is not surprising that the larger the organism as a whole, the larger the brain, but the human brain is out of line in this respect. It is larger than those of mammals far more massive than humans are. The human brain is larger than that of the horse, the rhinoceros, or the gorilla, for instance.

And yet the human brain is not the largest there is. The brains of elephants are larger. The largest elephant brains have been found to have masses of about 6 kilograms, or roughly 4¼ times that of the human brain. What's more, the brains of the large whales have been found to be more massive still. The most massive brain ever measured was that of a sperm whale, and it had a mass of about 9.2 kilograms—6½ times that of the human brain.

Yet elephants and large whales, while more intelligent than most animals, are never thought to compare even remotely with human beings in intelligence. Quite clearly, brain mass is not all there is to be considered where intelligence is concerned.

The human brain makes up about 2 percent of the mass of the total human body. An elephant with a 6-kilogram brain, however, would have a mass of 5,000 kilograms, so that his brain would make up only about 0.12 percent of the mass of its body. As for a sperm whale, which can attain a mass of 65,000 kilograms, its 9.2-kilogram brain would make up only about 0.014 percent of the mass of its body.

In other words, per unit of body mass the human brain is 17 times as large as that of the elephant, and 140 times as large as that of the sperm whale.

Is it fair to put brain/body mass ratio ahead of mere brain mass?

Well, it seems to give us a truthful answer, since it points up the apparently obvious fact that human beings are more intelligent than the larger-brained elephants and whales. Besides, we might argue it out (probably in a simplistic manner) in this fashion—

The brain controls the workings of the body, and what is left over from these low-thought-control duties can be reserved for activities such as imagination, abstract reasoning, and creative fancy. Though the brains of elephants and whales are large, the bodies of those mammals are enormous, so that their brains, large though they are, are fully preoccupied with all the routine of running those vast masses, and have very little left over for "higher" functions. Elephants and whales are therefore less intelligent than human beings despite the size of their brains.

(And that's why women can have brains 10 percent less massive than those of men, on the average, and not be 10 percent less intelligent. Their bodies are smaller, too, and their brain/body mass ratios are, if anything, a trifle higher than those of men.)

Still, the brain/body mass ratio can't be everything either. Primates (the apes and monkeys) all have high brain/body mass ratios and, on the whole, the smaller the primate, the higher the ratio. In some small monkeys the brain makes up 5.7 percent of the body mass, and that is nearly three times the ratio in human beings.

Why, then, aren't these small monkeys more intelligent than human beings?—Here the answer may be that their brains are just too small to serve the purpose.

MORE THINKING ABOUT THINKING

For really high intelligence, you need a brain massive enough to provide the thought power necessary, and a body small enough not to use up the entire brain and leave nothing for thinking. This combination of large brain and small body seems to meet its best balance in the human being.

But wait! Just as primates tend to have a higher brain/body ratio as they grow smaller, so do the cetaceans (the whale family). The common dolphin is no more massive than a man, on the whole, but it has a brain that is about 1.7 kilograms in mass, or ⅕ more massive than the human brain. The brain/body ratio is 2.4 percent.

In that case, why isn't the dolphin more intelligent than the human being? Can there be some qualitative difference between the two kinds of brains that condemns the dolphin to relative stupidity?

For instance, the actual brain cells are located at the surface of the cerebrum and make up the "gray matter." The interior of the brain is made up, to a large extent, of the fat-swathed processes extending from the cells and (thanks to the color of the fat) this is the "white matter."

It is the gray matter that is associated with intelligence, and therefore the surface area of the brain is more important than its mass. As we consider species in order of increasing intelligence, we find that the surface area of the brain increases more rapidly than the mass does. One way this becomes apparent is that the surface area increases to the point where it cannot be spread out smoothly over the brain's interior, but wrinkles into convolutions. A convoluted brain would have a larger surface area than a smooth brain of the same mass.

Therefore we associate convolutions with intelligence and, to be sure, mammalian brains are convoluted while nonmammalian brains are not. A monkey's brain is more convoluted than a cat's brain. A human brain, not surprisingly, is more convoluted than that of any other land mammal, even including such relatively intelligent ones as chimpanzees and elephants.

And yet the dolphin's brain is more massive than the human brain, has a higher brain/body mass ratio, and *in addition* is more convoluted than the human brain is.

Now why aren't dolphins more intelligent than human beings? To explain that, we have to fall back on the supposition that there is some shortcoming in the structure of the dolphin's brain cells, or in its cerebral organization—points for which there is no evidence.

Let me, however, suggest an alternative view. How do we know dolphins are not more intelligent than human beings?

To be sure, they don't have a technology, but that's not surprising. They live in water where fire is impossible, and the skillful use of fire is the fundamental basis of human technology. What's more, life in water makes streamlining essential, so that dolphins lack the equivalent of the human being's delicately manipulative hands.

But is technology alone a sufficient measure of intelligence? When it suits us, we dismiss technology. Consider the structures built by such social insects as bees, ants, and termites, or the delicate tracery of the spider web. Do these accomplishments make a bee, ant, termite, or spider more intelligent than the gorilla, which builds a crude tree nest?

We say "No" without a moment's hesitation. We consider that the lower animals, however marvelous their accomplishments, proceed only on instinct, and that this is inferior to conscious thought. Yet that may be only our self-serving judgment.

Might it not be conceivable that dolphins would consider our technology the result of a lower form of thinking, and dismiss it as evidence of intelligence, in a self-serving judgment of their own?

Of course, human beings have the power of speech. We

make use of complex modulations of sound to express infinitely subtle ideas, and no other species of living thing does that or comes anywhere near it. (Nor can they communicate with equivalent complexity, versatility, and subtlety by any other means, as far as we know.)

Yet the humpback whale sings complex "songs," while the dolphin is capable of producing a greater variety of different sounds than we can. What makes us so sure that dolphins can't or don't speak?

But intelligence is such a noticeable thing. If dolphins are so smart, why isn't it *obvious* that they are so smart?

I maintained in "Thinking About Thinking" that there are different kinds of intelligence among human beings and that IQ tests are misleading for that reason. Even if this were so, however, all the human intelligential (I have to invent that word) varieties clearly belong to the same genus. It is possible for us to recognize such varieties even though they are quite different. We can see that Beethoven had one kind of intelligence, Shakespeare another, Newton still another, and Peter Piper (the pickle-picking expert) yet another, and we can understand the value of each.

Yet what of an intelligential variety altogether different from anything any human being has? Would we even recognize it as intelligence at all, no matter how we studied it?

Suppose a dolphin, with its enormous, convoluted brain and its vast armory of sounds, had a mind that could consider complex ideas and a language that could express them with infinite subtlety. But suppose those ideas and that language were so different from anything to which we were accustomed that we could not even grasp the fact that they were ideas and language, let alone understand their content.

Suppose a colony of termites, all together, had a community brain that could react in a way so different from our individual ones that we would not see the community intelligence no matter how glaringly "obvious" it might be.

The trouble may be partially semantic. We insist on defining "thinking" in such a way that we come to the automatic conclusion that only human beings think. (In fact, bigots throughout history have been certain that only males rather similar in appearance to themselves could think, and that women and "inferior races" could not. Self-serving definitions can do a great deal.)

Suppose we defined "thinking" as being that sort of action that led to a species taking those measures that would best ensure its own survival. By that definition, every species thinks in some fashion. Human thinking becomes but one more variety, and one that is not necessarily better than others.

In fact, if we consider that the human species, with the full capacity for forethought, and knowing exactly what it is doing and what may happen, nevertheless has a very good chance of destroying itself in a nuclear holocaust—the only logical conclusion we can come to, by my definition, is that *Homo sapiens* thinks more poorly, and is less intelligent, than any species that lives, or has ever lived, on Earth.

It may be, then, that just as the IQ-niks achieve their results by carefully defining intelligence in such a way as to make themselves and people like themselves "superior," so humanity as a whole does the same by its careful definition of what constitutes thinking.

To make this plainer, let's consider an analogy.

Human beings "walk." They do so on two legs with their mammalian body tipped upward so as to produce a backward bend to the spine in the lumbar region.

We might define "walking" as motion on two legs with the body balanced on a recurved spine. By this definition, walking would be unique to the human being and we might well be proud of this fact, and with reason. This sort of walking freed our forelimbs from all necessity to help us move about (except under certain emergency conditions) and gave us permanently available hands. This development of upright posture preceded the development of our large brain and may, indeed, have led to it.

Other animals don't walk. They move on four legs or on six, eight, dozens, or none. Or they fly, or swim. Even those quadrupeds who can rise to their hind legs (such as bears and apes) do so only temporarily and are most comfortable on all fours.

There are animals that are strictly bipedal, such as kangaroos and birds, but they often hop rather than walk. Even birds that walk (as pigeons and penguins do) are primarily fliers or swimmers. And birds that never do anything but walk (or—its faster cousin—run), such as the ostrich, still lack the recurved spine.

Suppose, then, that we insisted on making "walking" totally unique and did so to the point where we lacked words for ways in which other species progressed. Suppose we were content to say that human beings were "walkative" and all other species were not, and refused to stretch our vocabulary beyond that.

If we insisted on doing so with sufficient fervor we would not need to pay any attention to the beautiful efficiency with which some species hop, or leap, or run, or fly, or sail, or dive, or slither. We would develop no phrase such as "animal locomotion" to cover all these varieties of progression.

And if we dismissed all forms of animal locomotion but our own as simply "non-walkative," we might never have to face the fact that human locomotion is, in many ways, not as graceful as that of a horse or a hawk, and is, indeed, one of the least graceful and admirable forms of animal locomotion.

Suppose, then, we invent a word to cover all the ways in which living things might behave in such a way as to meet a challenge or to promote survival. Call it "zorking." Thinking, in the human sense, might be one form of zorking, while other species of living things might display other forms of zorking.

If we approach zorking without preconceived judgments, we might find that thinking is not always the best way of zorking, and we might stand a slightly better chance of understanding the zorking of dolphins or of termite communities.

Or suppose we consider the problem of whether machines can think; whether a computer can ever have consciousness; whether robots can possibly feel emotion; where, in short, we are to attain, in future, such a thing as true "artificial intelligence."

How can we argue such a thing, if we don't first stop to consider what intelligence might be? If it is something only a human being can have by definition, then, of course, a machine can't have it.

But any species can zork, and it may be that computers will be able to, also. Computers won't zork, perhaps, in a fashion that any biological species will, so we need a new word for what they do, too. In my impromptu talk at the computer force, I used the word "grotch" and I suppose that will do as well as any other.

Among human beings there are an indefinite number of different ways of zorking; different ways that are sufficiently alike for all to be included under the general heading of "thinking." And among computers, too, there are liable to be an indefinite number of different ways of zorking, but ways so different from those found in human beings as to be included under the general heading of "grotching."

(And nonhuman animals may zork in different ways still, so that we would have to invent dozens of different words for zork varieties and classify them in complicated fashion. What's more, as computers developed, we might find that "grotching" wasn't sufficient, so that we would have to work up subheadings.—But all that's for the future. My crystal ball isn't infinitely clear.)

To be sure, we design our computers in such a way that they can solve problems that are of interest to us, and therefore they give us the illusion that they think. We must recognize, though, that even when a computer solves a problem that we ourselves would have to solve in the absence of a computer, it and we nevertheless solve it by totally different processes. They grotch and we think, and it may be useless to sit about and debate whether computers think. Computers might as well sit around and debate whether human beings grotch.

But is it reasonable to suppose that human beings would create an artificial intelligence so different from human intelligence as to require a recognition of computer grotching as independent of human thinking?

Why not? It's happened before. For countless thousands of years, human beings transported objects by tucking them under their arms or balancing them on their heads. In doing so, they could only transport so much mass, at most.

If human beings piled objects on the backs of donkeys, horses, oxen, camels, or elephants, they could transport larger masses. That, however, is just the substitution of the direct use of larger muscles for smaller ones.

Eventually, however, human beings invented an artificial device that made transportation easier. How did the machine bring this about? Did it do it by producing an artificial walk, run, fly, swim, or any of the myriad of other forms of animal locomotion?

No. Some human being in the dim days of prehistory invented the wheel and axle. As a result, a much larger mass could be placed in a cart and dragged by human or animal muscles than could be carried by those muscles directly.

The wheel and axle is the most astonishing invention ever made by human beings, to my way of thinking. The human use of fire was at least preceded by the observation of natural fires set by lightning. But the wheel and axle had no natural ancestry. It does not exist in nature; no life-form has evolved it to this day. Thus, "machine-aided locomotion" was, from its inception, utterly different from all forms of animal locomotion; and, in the same way, it would not be surprising if mechanical zorking were different from all forms of biological zorking.

Of course, primitive carts couldn't move by themselves, but eventually the steam engine was invented, and then later the internal combustion engine and the rocket—none of which acts anything like muscles.

Computers are, as yet, at the pre-steam-engine stage. Computers can work but can't do so "by themselves." Eventually the equivalent of a steam engine will be developed and the computers will be able to solve problems by themselves, but still by a process totally different from that of the human brain. They will still be grotching rather than thinking.

All this seems to rule out fears that computers will "replace" us, or that human beings will become superfluous and die out.

After all, wheels haven't made legs superfluous. There are times when walking is more convenient and more useful than rolling. Picking one's way over rough ground is easy when walking, very difficult by automobile. And I wouldn't think of getting from my bed to the bathroom by any process other than walking.

But can't computers eventually do anything human beings

can do, even if they grotch rather than think? Can't computers grotch up symphonies, dramas, scientific theories, love affairs—anything you care to name?

Maybe. Every once in a while I see a machine designed to lift legs over obstructions so that it walks. However, the machine is so complicated and the motions so ungraceful that it strikes me that no one is ever going to take the huge trouble to try to produce and use such things as anything but a tour de force (like the airplane that flew the English Channel by bicycle power—and will never be used again).

Grotching, whatever it is, is quite obviously best adapted to the exceedingly rapid and inerrant manipulation of arithmetical quantities. Even the simplest computer can grotch the multiplication and division of huge numbers much faster than human beings can think their way through to the solution.

That doesn't mean grotching is superior to thinking; it just means that grotching is better adapted to that particular process. As for thinking, that is well adapted to processes that involve insight, intuition, and the creative combination of data to produce unexpected results.

Computers can perhaps be designed to do such things after a fashion, just as mathematical prodigies can grotch after a fashion—but either is a waste of time.

Let thinkers and grotchers work at their specialties and pool the results. I imagine that human beings *and* computers, working together, can do far more than either could alone. It is the symbiosis of the two that represents the shape of the future.

One more point: If grotching and thinking are widely different things, can one expect the study of computers ever to illuminate the problem of human thought?

Let us go back to the problem of locomotion.

A steam engine can power machines to do work that is

ordinarily done by muscles, and to do it more intensely and tirelessly, but that steam engine is in no way similar to the muscle in structure. In the steam engine, water is heated to steam and the expanding steam pushes pistons. In the muscle, a delicate protein named actomyosin undergoes molecular changes which cause the muscle to contract.

It might seem that you could study boiling water and expanding steam for a million years and not be able to deduce one thing about actomyosin from that. Or, conversely, that you could study every molecular change actomyosin can undergo and not learn one thing about what makes water boil.

In 1824, however, a young French physicist, Nicolas L. S. Carnot (1796–1832) studied the steam engine in order to determine what factors controlled the efficiency with which it worked. In the process, he was the first to begin the line of argument that, by the end of the century, had fully developed the laws of thermodynamics.

These laws are among the most powerful generalizations in physics and it was found that they applied in full rigor to living systems as well as to such simpler things as steam engines.

Muscular action, however complicated its innermost workings, must labor under the constraints of the laws of thermodynamics just as steam engines must, and this tells us something about muscles that is enormously important. What's more we learned it from steam engines, and might never have learned it from a study of muscles alone.

Similarly, the study of computers may never tell us, directly, anything at all about the intimate structure of the human brain, or of the human brain cell. Nevertheless, the study of grotching may lead to the determination of the basic laws of zorking, and we may find that these laws of zorking apply to thinking as well as to grotching.

MORE THINKING ABOUT THINKING

It may be, then, that even though computers are nothing like brains, computers may teach us things about brains that we might never discover by studying brains alone—so, in the last analysis, I am on Minsky's side.

12

COMING FULL CIRCLE

During the autumn of 1983 I grew fascinated by the currently very popular bypass operations, and for a very good reason, too. My angina, which had been minor and stable for six years, had suddenly flared up. There were tests, and once I was carefully exposed to the results of those tests, I realized I had those most interesting of alternatives—none.

I was going to need a triple bypass.

So I inquired of my various doctors and there seemed to be one question they didn't hear me ask. At least they always launched into some other answer.

Finally, I cornered my anesthesiologist.

I said, "There is one thing I don't understand. If you're going to insert an artery or vein into my aorta and into my coronary arteries so that the blood will be led around the constriction, how do you do that? Short of making use of the fourth dimension, you're going to have to cut into the aorta, for instance, and make a round hole into which you can fit the new vessel."

"Well, yes."

"And at the first cut," I said, "the blood gushes out in a tremendous flood and I die."

"Oh, *no*," he said. "Hasn't anyone explained? After we expose your heart, we *stop* it."

I felt myself turn a pretty shade of green. "You *stop* it?"

"Yes, we give it a heavy slug of potassium ion and cool it and it stops beating.

"But that leaves me five minutes from brain death."

"No, it doesn't. You will be fitted to a heart-lung machine that will keep you going for hours, if necessary."

"What if it breaks down?"

"It won't. And if there's a power breakdown all over the Northeast, we continue untouched on our emergency generators."

I felt a *little* better, and said, "How do you start the heart again? What if it won't start?"

"That can't happen," he said confidently. "The heart wants to do nothing but start. We have to work hard to keep it still. As soon as we let the potassium wash out, it starts right in again, especially one that's in as good shape as yours."

He was right. The triple bypass took place on 14 December 1983, and on 2 January 1984 I celebrated my 1,000,000th birthday (on the binary scale) and here, on 8 January 1984, I begin another essay.—And what ought I to discuss but the heart and blood vessels?

Aristotle (384–322 B.C.) thought the heart was the seat of the intelligence. That wasn't as unreasonable as it sounds today. After all, it is a constantly moving organ that speeds up during excitement, slows down in periods of calm, is tumultuous when one's affections are engaged and so on. Anyone observing this and then noting that the brain just sits there and does nothing is likely to dismiss the brain as, at best, an auxiliary organ.

Aristotle thought it was merely a cooling agent for the heart, which would otherwise overheat. The cooling was done by a salivalike fluid which the Greeks called *pituita* (the first syllable of which gives us our onomatopoetic word "spit"). There is still a small organ at the base of the brain called the pituitary gland, which is extremely important (another essay, someday, perhaps) but which has nothing to do with spit.

Aristotle did not distinguish between veins, arteries, nerves, and sinews.

Soon after Aristotle's death, however, there came a short period of clever dissection in Alexandria, Egypt, and matters began to straighten out somewhat.

The arteries, for instance, were clearly connected to the heart, but in dead bodies the large arteries seemed empty. (The last pulsations cleared them of blood.) Praxagoras (340– ? B.C.) made the logical suggestion, therefore, that they carried air. In fact, the word "artery" is from a Greek term meaning air duct.

Herophilus (320– ? B.C.), a student of Praxagoras, noted that the arteries pulsed and the veins did not. He seems to have believed the arteries carried blood, but he kept the name his teacher had given them.

Herophilus's disciple, Erasistratus (304-250 B.C.), believed that the veins, arteries, and nerves were all hollow tubes that carried one fluid or another to the various parts of the body; that they divided and subdivided till they were too small to see. In all this he was remarkably close, since even the nerves carry an electric impulse, which can be looked upon as a subtle sort of fluid.

The veins, all agreed, carried blood. ("Vein" is from the

COMING FULL CIRCLE

Latin *vena*. The Greek word is *phleb*, which is why inflammation of the veins is called phlebitis.)

Some thought the arteries contained a mixture of blood and air, or of blood and some "vital spirit," and if we think of the arteries as carrying oxygenated blood, which they do, we find the old Greeks were not making unintelligent guesses.

Still, puzzles remained, and it was centuries before medical men clearly understood that nerves and tendons had nothing to do with the heart and were not blood vessels of any type. Nor did they get the difference between veins and arteries entirely clear.

Galen, the most renowned of ancient physicians, a Greek of Roman times (130-200), thought the arteries originated in the heart and led to the various tissues. He thought the veins originated in the liver, went from there to the heart from which, again, they were led to the various tissues. (Not an unintelligent guess, actually. The liver is a large organ that is full of blood vessels, and, whereas the arteries beat as the heart does, the veins remain quiescent as the liver does.)

Galen believed that the blood poured out of the heart through arteries and veins alike and was consumed by the tissues. New blood was continually manufactured, he thought, in the liver (presumably out of food), as fast as it was consumed in the tissues. The blood was consumed in the tissues as wood would be in a fireplace. The air we breathed stoked the process, and the air we exhaled was analogous to the smoke of a fire.

There is a catch, though. The heart is not a simple pump. It is in fact two pumps, since it is divided into two chief chambers: the left ventricle and the right ventricle. ("Ventricle" is from the Latin for "little pouch.")

Each thick-walled ventricle has a thin-walled antechamber, called the "left atrium" and "right atrium" respectively, so the heart has four chambers altogether.

There is a clear passage between each atrium and its ventricle but none between the two sets of atria-ventricles. The left ventricle (very muscular) leads to the largest artery in the body, the aorta (a name of uncertain origin) while the right ventricle (less muscular) leads to the pulmonary artery. Each ventricle has its own veins, too.

It would seem that each ventricle sends out blood and that there is no obvious connection between the two bloodstreams. Galen, however, could not see why there should be two bloodstreams and decided that made no sense. There must be a connection, and if not an obvious one then a hidden one.

The wall between the two ventricles is thick and muscular and, to all appearances, absolutely intact. Nevertheless, Galen reasoned, there must be tiny holes in it, holes too small to see, through which blood was transferred back and forth from one ventricle to the other, thus allowing there to be a single bloodstream.

For something like fourteen centuries physicians faithfully believed in the interventricular pores, although no one saw them, and though they do not, in actual fact, exist.—But don't be too ready to laugh. They made sense in Galen's system, and even though the system proved wrong, the correct system, when it came along, also depended on invisible passages.

There was no chance of making progress, however, on the matter of the heart and blood vessels until human anatomy was established as a firm medical discipline. This was difficult because so many people considered the dissection of corpses (I'm not talking about the vivisection of living bodies) to be blasphemous. The Egyptians, Jews, and, eventually, Christians, were alike horrified at the practice, and anatomy died out after 200 B.C. and was restricted to animals for a thousand years.

COMING FULL CIRCLE

The first great modern medical schools in Europe were established in Renaissance Italy and they led the Western world for three centuries. At the University of Bologna, Mondino de'Luzzi (1275–1326) was the first to conduct systematic dissections. In 1316 he published the first book in history to be devoted entirely to anatomy. Unfortunately he had assistants do the dissection, while he did the lecturing (without looking) from Galenic principles. He made egregious errors, therefore, but for two and a half centuries his remained the best book available.

(Incidentally, the growth of interest in naturalistic art in Renaissance Italy made anatomy an artistic necessity; just as it made projective geometry one. Art thus contributed to medicine and mathematics, while each of these in turn contributed to art. There are intellectual and technological connections everywhere in history. Leonardo da Vinci—1452–1519—dissected thirty cadavers in the course of his life.)

Finally, there came along the first great modern anatomist, a Fleming named Andreas Vesalius (1514–64). He studied in the Italian medical schools and grew fascinated by anatomy. He managed to create a sensation, for instance, by showing that men and women had equal numbers of ribs—twenty-four each, arranged in twelve pairs.

After all, the Bible said that Eve was created out of a rib taken from Adam—from which it followed that there had to be one missing not only in Adam, but in all men. Everyone "knew" that without having to look—until Vesalius went and looked and, what was worse, counted.

As a result of his researches, Vesalius put together one of the great classics of scientific history, entitled, in English, *On the Structure of the Human Body*. It was published in 1543 when he was twenty-nine years old, and that was the same year in which Copernicus published his book explaining that the Earth went round the Sun and not vice versa. It was a double whammy for Greek science.

Vesalius's book was the first reasonably accurate book on anatomy, and it was *printed*. This meant it could have illustrations that could be reproduced accurately any number of times, and Vesalius managed to get a first-class artist to do them, one Jan Stephen van Calcar (1499–1550), a pupil of the artist Titian (1477–1576). The illustrations were naturalistic, and those of the muscles, in particular, have never been done better.

Older, more conservative anatomists objected strenuously to the book, simply because they couldn't let go of Galen. Twenty years later, they managed to help get Vesalius accused of heresy, body-snatching and dissection. He was forced to go on a pilgrimage to the Holy Land as penance, and died in the course of a storm.

However, even Vesalius did not abandon Galen altogether. He was for Galen and against Aristotle in favoring the brain over the heart as the seat of the intelligence—and no one has doubted the matter since.

Furthermore, Vesalius could find no way, in his anatomical researches, of explaining the double-pump nature of the heart other than the way Galen did. He therefore accepted the invisible pores through the interventricular wall of the heart, though toward the end of his life he is supposed to have grown doubtful about it.

Despite Vesalius's troubles with the establishment of his day, he revolutionized anatomy. After him, anatomists dissected carefully and studied in detail what they saw.

One of these was Girolamo Fabrici, 1537–1619 (usually better known as Fabricius ab Aquapendente.) In 1574 he studied the leg veins and noted that they had little valves along their length. Other anatomists of the time also reported on them and there were vigorous disputations over priority.

Fabrici, however, did the most careful and thoroughgoing study and allowed one of his students to publish illustrations of these valves in 1585, so Fabrici usually gets the credit for the discovery.

Fabrici, however, did not interpret their function correctly. He was still enslaved to the Galenic notion of interventricular pores that allowed a single bloodstream to move out centrifugally from the heart to the tissues, where it was consumed.

It was clear that the valves prevented the blood from flowing downward in the veins. Muscular action, during walking and other movements, squeezed the leg veins and other veins in the lower body and forced the blood upward because that was the only direction in which it could go. If it tried to go downward in the direction of gravitational pull, the valves caught it.

This meant that the blood in the leg veins and, possibly, in all veins could move *toward the heart only*.

Fabrici, however, couldn't accept that despite the fact that (in hindsight) it is plainly so. He assumed that the valves merely slowed and equalized the downward flow of blood so that all parts of the body got their fair share. In doing this, Fabrici saved the Galenic theory of heart action but lost immortality.

Did no one question the Galenic pores?

Some did certainly, but the first to do so was not a European. He was an Arabic scholar, Ibn al-Nafis (1210–88), born near Damascus.

In 1242 he wrote a book on surgery, and in it he specifically denied the existence of the Galenic pores. The interventricular wall, he said, was thick and solid and there was no way for blood to get through it.

And yet blood had to get from one side of the wall to the other somehow. A double pump made no sense.

Al-Nafis suggested that the blood from the right ventricle was pumped into the pulmonary artery that led it to the lungs. There, in the lungs, it divided into smaller and smaller vessels, within which the blood picked up air from the lungs. The vessels were then collected into larger and larger vessels until they emptied into the pulmonary veins that carried the blood, together with its air admixture, into the left atrium and from that into the left ventricle and out the aorta.

Al-Nafis, in this way, discovered "the lesser circulation" of the blood, and the picture was an interesting one. The blood (created in the liver, perhaps, as Galen thought) poured into the right atrium and right ventricle, then traveled to the left atrium and left ventricle by way of the lungs. Then, aerated, the blood traveled to the tissues generally.

In this way you got rid of the Galenic pores and explained the reason for the double pump. It was a way of ensuring that the blood picked up air before going to the tissues generally.

There were two catches to al-Nafis's theories, however. In the first place, there were no signs of continuous blood vessels through the lungs. The pulmonary artery divided and subdivided till it disappeared, while the pulmonary veins formed out of apparent nothingness. It was fair to suppose that the final subdivisions grew too small to see and that the tiniest arteries and veins were thus connected. However, in that case, invisible vessels replaced invisible pores. Was that really an improvement?

The second catch was that al-Nafis's book did not become known to the West until 1924 (!), and therefore had no influence on the development of modern medical theory.

It took a little over three hundred years for Europe to catch up to al-Nafis's insight, and the one who did it was a Spanish physician, Michael Servetus (1511–53).

Those were the days of the Protestant Reformation and all Europe was convulsed with theological discussions. Servetus developed radical notions that would today be described as Unitarian. He advanced them tactlessly, infuriating both Catholics and Protestants, since both were committed to the divinity of Jesus. In 1536 Servetus met John Calvin in Paris. John Calvin was one of the most noted of the early Protestants, a firm and dour doctrinaire. When Servetus sent Calvin a copy of his views, Calvin was horrified and broke off the correspondence—but did not forget.

In 1553 Servetus published his theological views anonymously, but Calvin knew those views and recognized the author. Calvin transmitted the knowledge to the French authorities, who arrested Servetus. Servetus managed to escape three days later and headed for Italy.

Foolishly, he went by way of nearby Geneva, then under the strict control of the dark and bitter Calvin, who had set up one of the more notable theocracies of modern Europe. Servetus was not a subject or resident of Geneva and had committed no crime in Geneva for which he could be legally held. Nevertheless, Calvin insisted on having him condemned to death, so that Servetus—crying out his Unitarian doctrine to the end—was burned at the stake.

Calvin was not satisfied with burning Servetus's body. It seemed to him necessary to burn his mind, too. He hunted down all the thousand copies of Servetus's book he could find and had them burned as well. It was not until 1694, a century and a half after Servetus's death, that some unburned copies were discovered and European scholars had a chance to read his Unitarian views.

This they did, and found to their astonishment, perhaps, that he had also described the lesser circulation in the book (exactly as al-Nafis had done, if Europe had only known).

Servetus lost the credit for the discovery, except in hindsight, for in 1559 an Italian anatomist, Realdo Colombo (1516–59), had published a book that described the lesser circulation just as al-Nafis and Servetus had done, and this work survived. Colombo generally gets the credit for the discovery, but then his work was more detailed and careful than that of the other two, and, through circumstance, it was Colombo's work that influenced further developments so his credit is deserved.

Then came the English physician William Harvey (1578–1657).

He was the son of a well-to-do businessman and the oldest of nine children. He received his degree at Cambridge in 1597, and then went to Italy to study medicine. There Fabrici was one of his teachers.

Harvey returned to England to great success, for he was court physician to both King James I and King Charles I.

Harvey was an experimenter. To him, the heart was a muscle that was ceaselessly contracting and pushing out blood, and it was to be investigated on that basis and no other.

By actual dissection, he studied the valves between the two atria and the two ventricles carefully and noticed that they were one-way. Blood could travel from the left atrium to the left ventricle and from the right atrium to the right ventricle, but not vice versa.

What's more, Harvey knew, of course, of the venous valves that his old teacher Fabrici had reported on. With the concept of one-way valves clearly in his mind, he avoided Fabrici's mistake. The blood in the veins went in one direction only, toward the heart. He experimented by actually tying off veins in the course of animal experiments. Inevitably the blood filled and bulged the vein on the side away from the heart as it tried to flow toward the heart and couldn't flow away. The situation was precisely reversed when he tied off an artery, which at once filled and bulged with blood on the side toward the heart.

COMING FULL CIRCLE

To Harvey, by 1615 matters were clear. He finally knew the physiological difference between arteries and veins. Blood left the heart by way of the arteries, and returned to the heart by way of the veins. The lesser circulation that Colombo had talked about was *only* the lesser. From the left ventricle, the blood was pumped out into the aorta and went to all the tissues of the body generally, returning by veins to the right atrium and ventricle, from which it was pumped out to the lungs to return to the left atrium and ventricle.

In other words, the blood is constantly coming full circle. It "circulates."

Harvey did some simple calculations that Galen might conceivably have done, if the notion of measurement in connection with biology had been clear to the Greeks. Harvey showed that in one hour the heart pumped out a quantity of blood that was three times the weight of a man. It seemed inconceivable that blood would be formed and consumed at that rate, so the notion of the circulation of the blood seemed a logical necessity as well as an experimental one.

Harvey, who was not a controversialist, began lecturing on the circulation of the blood in 1616, but didn't put it into book form until 1628. He then produced a seventy-two-page book, miserably printed in the Netherlands on thin, cheap paper, and full of typographical errors. However, the experiments it described were clear, concise, and elegant and the conclusions were incontrovertible. The book (called, in English, *On the Motions of the Heart and Blood*) therefore became one of the great scientific classics.

Harvey's book was (inevitably) hooted down at first, but he lived long enough to see the circulation of the blood accepted by European medicine generally. It was his book that was the final once-and-for-all ending of Galenic physiology.

—And yet there was a catch. The blood was carried outward from the heart by the arteries and carried back by the veins, but there were no visible connections between the two. One had to assume the existence of invisible connections— little tubes too small to see, like the invisible Galenic pores in the interventricular muscle.

As long as we must depend on invisibility, we are not certain.

Ah, but now there was a difference. In Harvey's last decade of life, physiologists were beginning to use microscopes that were very imperfect but that could magnify tiny objects ordinarily too small to see and make them visible in some detail.

First in the field by a little was the Italian physiologist Marcello Malpighi (1628–94), who had been trained in medicine at the University of Bologna and who eventually rose to become, rather reluctantly, private physician to Pope Innocent XII.

Malpighi began his work in microscopy in the 1650s when he investigated the lungs of frogs. He began to see tiny hairlike blood vessels he couldn't see without the microscope. By observing the wing membranes of bats under his microscope in 1661, he could actually see tiny arteries and tiny veins connected by such hairlike vessels. He called them capillaries, from Latin words meaning "hairlike."

The discovery, which completed and made perfect the concept of the circulation of the blood, was made four years (alas) after Harvey's death, but I'm quite sure Harvey was confident that capillaries existed and would be discovered.

One last item. When the heart's left ventricle pumps its blood into the huge aorta, three small arteries come off it almost immediately, taking the most freshly oxygenated blood to—where else?—the heart muscle itself. The heart helps itself first and most richly, and why not? It deserves it.

These vessels are the "coronary arteries" (because they encircle the heart like coronets). Even more so than ordinary arteries, the coronaries have a tendency to clog with cholesterol, if one eats and lives foolishly.

The clogging is usually just at the place where the arteries branch off from the aorta, and tests showed that my coronaries (in order of decreasing size) were clogged 85 percent, 70 percent, and 100 percent.

The largest coronary was bypassed by a nearby artery (fortunately perfectly usable). The two smaller ones were bypassed by a vein taken from my left leg.

I am scarred and not yet totally healed, but my heart is getting all the blood it needs; I am healing rapidly; and—by far the most important—I can still write these chapters.

PART V
TECHNOLOGY

13

WHAT TRUCK?

I am not a very visual person. What's more, I have a very
lively inner life so that things are jumping about inside my
cranium all the time and that distracts me. Other people are
therefore astonished at the things I don't see. People change
their hairstyles and I don't notice. New furniture comes into
the house and I sit on it without comment.

Once, however, I seem to have broken the record in
that respect. I was walking up Lexington Avenue, speaking animatedly (as is my wont) to someone who was walking with me. I walked across a roadway, still talking, my
companion crossing with me with what seemed to be a certain
reluctance.

On the other side, my companion said, "That truck missed
us by one inch."

And I said, in all innocence, "What truck?"

So I got a rather long dull lecture, which didn't reform me,
but which got me to thinking about the ease with which one
can fail to see trucks. For instance . . .

TECHNOLOGY

* * *

Some time ago a reader sent me a copy of the October 1903 issue of *Munsey's Magazine* and I looked through it with considerable interest. The enormous advertising section was like a window into another world. The item of particular fascination, however, which the reader had called to my attention was an article entitled "Can Men Visit the Moon?" by Ernest Green Dodge, A.M.

It was the kind of article I myself might have written eighty years ago.

As it happens, I have often had occasion to wonder whether my own attempts to write about the technology of the future might seem less than inspired in the brilliant light of hindsight. I have usually felt, rather woefully, that it would—that it would turn out that there would be trucks I didn't see, or trucks I saw that weren't really there.

I can't expect to live eighty more years and check on myself, but what if I look at remarks I might have made eighty years ago and see how they would sound in the light of what we now know?

Mr. Dodge's article is the perfect way of doing this, for he was clearly a rational man with a good knowledge of science, and with a strong but disciplined imagination. In short, he was as I like to imagine that I am.

In some places he hits the target right in the bulls-eye.

Concerning a trip to the Moon, he says: ". . . it is not, like perpetual motion or squaring the circle, a logical impossibility. The worst that can be said is that it now looks as difficult to us as the crossing of the great Atlantic must once have appeared to the naked savage upon its shore, with no craft but a fallen tree and no paddle but his empty hands. The impossibility of the savage became the triumph of Columbus, and the day-dream of the nineteenth century may become the achievement even of the twentieth."

Exactly! Human beings were standing on the Moon only sixty-six years after Dodge's article had appeared.

Dodge goes on to list the difficulties of space travel, which, he points out, arise primarily out of the fact that "space is indeed empty, in a sense which no man-made vacuum can approach. . . . a portion of outer space the size of the earth contains absolutely nothing, so far as we know, but a few flying grains of meteoric stone, weighing perhaps ten or fifteen pounds in all."

Dodge is a careful man. Although the statement seemed irrefutable in 1903, he inserts that cautious phrase "so far as we know" and was right to do so.

In 1903 subatomic particles were just becoming known. Electrons and radioactive radiations had been discovered less than a decade before. These were only earthbound phenomena, however, and cosmic rays were not discovered until 1911. Dodge could not, therefore, have known that space was filled with energetic electrically charged particles of insignificant mass, but considerable importance.

On the basis of what he did know in 1903, Dodge lists four difficulties that could arise in traveling from Earth to Moon through the vacuum of outer space.

The first is, of course, that there is nothing to breathe. He dismisses that, quite correctly, by pointing out that a spaceship would be airtight and would carry its own internal atmosphere, just as it would bring along supplies of food and drink. Breathing is therefore no problem.

The second difficulty is that of "the terrible cold" of outer space. This Dodge takes more seriously.

It is, however, a problem that tends to be overestimated. To be sure, any piece of matter that is in deep space and far from any source of radiation would reach an equilibrium temperature of about 3 degrees absolute, so this can be viewed as "the temperature of space." Anything traveling from the

Earth to the Moon is, however, not far from a source of radiation. It is in the vicinity of the Sun, just as much as the Earth and Moon are, and is bathed in solar radiation all the way.

What's more, the vacuum of space is an excellent heat insulator. This was well known in 1903, for James Dewar had invented the equivalent of the thermos bottle eleven years before the article was written. There is sure to be internal heat within the spaceship, if only the body heat of the astronauts themselves, and this will be lost very slowly by way of radiation through vacuum. (This is the only way of losing heat in space.)

Dodge thinks that the ships would have to be guarded against heat loss by "having the walls . . . heavily padded." He also suggests a heat supply in the form of "large parabolic mirrors outside [which] would throw concentrated beams of sunlight through the window."

This is a sizable overestimate, since nothing of the sort is necessary. Insulation must be placed on the outside of the ships, but that is for the purpose of avoiding the *gain* of too much heat during passage through the atmosphere. The *loss* of heat is of no concern to anyone.

The third difficulty arises from the fact that the ship will be in free fall during most or all the passage from the Earth to the Moon so that the astronauts will experience no gravitational pull. This Dodge very reasonably shrugs off, pointing out that "dishes could be fastened to the table, and people could leap and float, even if they could not walk."

He does not speculate on possible deleterious physiological changes arising from exposure to zero gravity, and this might be considered shortsighted. Still, this has turned out not to be a problem. In recent years people have remained under zero gravity conditions continuously for more than half a year and have apparently shown no permanent ill-effects.

The fourth and last danger that Dodge considers is that of the possibility of meteoric collisions but (despite the fact that science fiction writers continued to view that as a major danger for half a century longer) Dodge dismisses this, too, as statistically insignificant. He was right to do so.

He does not mention the fifth danger, that of the cosmic rays and other electrically charged particles, something he simply could not have known about in 1903. There were some misgivings about this after the discovery of the radiation belts in 1958 but, as it turns out, they did not materially interfere with humanity's reach to the Moon.

Dodge thus decided that there were no dangers in space that would prevent human beings from reaching the Moon, and he was right. If anything, he had overestimated the danger of the supposed cold of space.

The next question was exactly how to go about actually traversing the distance from Earth to Moon. In this connection, he mentions five possible "plans." (One gets the impression, though Dodge does not actually say so, that these five plans are the only ones that are conceivable.)

The simplest is "the Tower Plan." This would involve the construction of an object tall enough to reach the Moon, something like the scheme of the builders of the biblical tower of Babel. Dodge mentions the Eiffel Tower, which had been built fourteen years before, and which, at a height of 984 feet, was the tallest structure in the world at the time the article was written (and for twenty-seven years longer).

He says: "the combined wealth of all nations might construct an edifice of solid steel eight or ten miles in height, but not much more, for the simple reason that the lower parts could not be made strong enough to bear the weight that must rest upon them." To reach the Moon, there would have to be "a building material about five hundred times tougher than armor-plate, and such may never be discovered." (Note the "may." Dodge is a careful man.)

There are many other deficiencies to the plan that Dodge does not mention. The Moon, having an elliptical orbit at an angle to Earth's equatorial plane, would approach the top of the tower only once in a long while, and when it did so, the lunar gravity would produce a huge strain on it. Air would remain only at the bottom of the tower, thanks to the pull of Earth's gravity, and there would still be the problem of traversing the 300,000 kilometers or so to the Moon's perigee distance after the tower had been built (let alone traversing it in the process of building the tower). Scratch the "Tower Plan."

Dodge doesn't mention the variant possibility of a "sky-hook," a long vertical structure in such a position between Earth and Moon that the combined gravitational pull holds it in place, and that can be used as a help in negotiating the Earth-Moon passage. Personally, I don't think that's any-where near practical, either.

Dodge's second scheme is "the Projectile Plan." This involves shooting a ship out of a giant cannon and having it emerge with speed enough to reach the Moon (if correctly aimed). This is the method used by Jules Verne in *From the Earth to the Moon*, published thirty-eight years before, in 1865.

Dodge points out that to reach the Moon, the projectile must leave the muzzle of the cannon at a speed of 11.2 kilometers per second (the escape velocity from Earth) plus a little extra to make up for air-resistance losses in passing through the atmosphere. The spaceship would have to accel-erate from rest to 11.2 kilometers per second in the length of the cannon barrel, and this would neatly crush all passengers on board, leaving not a bone unbroken.

The longer the cannon, the lower the acceleration, but,

says Dodge, ". . . even if the gun barrel had the impossible length of forty miles, the poor passenger would be subjected for eleven seconds to a pressure equivalent to a hundred men lying upon him."

But suppose we could overcome that difficulty and somehow picture the spaceship leaving the cannon's mouth with the passengers on it still alive. The spaceship would then be a projectile, moving in response to the force of gravity and nothing more. It would be as unable to alter its course as any other cannonball would.

If the ship were aimed at the Moon and were eventually to land on it, it would have to strike with a speed of not less than 2.37 kilometers per second (the Moon's escape velocity). And that, of course, would mean instant death. Or, as Dodge says, ". . . unless our bullet-ship can carry on its nose a pile of cushions two miles high on which to light, the landing will be worse than the starting!"

Of course, the ship need not land on the Moon. Dodge does not pursue this plan farther, but the cannon may be aimed with such superhuman nicety as to miss the Moon by just the right amount and at just the right speed to cause it to move around the Moon in response to the lunar gravity and race back to a rendezvous with Earth.

If the ship then hit the Earth squarely, it would strike at no less than 11.2 kilometers per second, so the passengers would be fried to death on the passage through the atmosphere before being blasted to death on collision with the solid ground or (very little better at such a speed) the ocean. And if the spaceship hit a city, it would kill many thousands of innocents as well.

The original superhuman aim might have brought the ship back to Earth just sufficiently off-center to trap it in Earth's gravity and put it into an orbital path within the upper reaches of Earth's atmosphere. The orbit would gradually decay.

Furthermore, some parachute arrangement might then be released to hasten the decay and bring the ship down safely.

But to expect all that of one aim is to expect far too much even if the initial acceleration were not murderous. Scratch the Projectile Plan.

The third scheme is "the Recoil Plan."

Dodge points out that a gun can fire in a vacuum and, in the process, undergo recoil. We can imagine a spaceship that is a kind of mighty gun which could eject a projectile downward, so that it would itself recoil upward. While recoiling, it could eject another projectile downward and give itself another kick upward.

If the ship fired projectiles rapidly enough, it would recoil upward faster and faster and, in fact, recoil itself all the way to the Moon.

Dodge, however, argues that the recoil is increasingly great as the mass of the bullet increases, and that "to be effective its weight [mass, really] should equal or exceed that of the gun itself."

We must imagine, then, an object that would fire away half of itself, leaving the other half to move upward—and fire half of what is left of itself as it rises, thus moving upward faster—and then fire half of what is now left of itself—and so on, until it reaches the Moon.

But how big must a spaceship be to begin with, if it has to fire away half of itself, then half of what is left, then half of what is left and so on? Dodge says, "An original outfit as big as a mountain chain would be necessary in order to land even a small cage safe upon the lunar surface." He feels that the Recoil Plan is therefore even less practical than the Projectile Plan.

On to the fourth scheme, "the Levitation Plan."

This involves nothing less than the screening, somehow, of

the force of gravity. Dodge admits that no such gravity-screen is known, but supposes that it might be possible to discover one at some time in the future.

In a way, a hydrogen-filled balloon seems to nullify gravity. Indeed, it seems to fall upward through the atmosphere and to exhibit levitation (from a Latin word meaning "light") rather than gravitation (from a Latin word meaning "heavy").

Edgar Allan Poe, in his story "The Unparalleled Adventure of One Hans Pfaall," published sixty-eight years before, in 1835, made use of a balloon to travel to the Moon. A balloon, however, merely floats on the denser layers of the atmosphere and does not truly neutralize gravity. When it rises to a height where the thinning atmosphere is no denser than the gas contained within the balloon, there is no further rise. Poe imagined a gas far less dense than hydrogen (something we now know does not, and cannot, exist) but even that could not have lifted a balloon more than a fraction of one percent of the distance to the Moon. Dodge knew that and did not so much as mention balloons.

What Dodge meant was a true gravity neutralization, such as H. G. Wells used in his story "The First Men in the Moon," published two years before, in 1901.

Of course if you neutralized gravity you would have zero weight, but would that alone carry you to the Moon? Would a spaceship with zero weight not merely be subject to the vagaries of every puff of air? Would it not simply drift this way and that in a sort of Brownian motion and even if, eventually (a long eventually, perhaps), it got to the top of the atmosphere and went beyond, might it not then be moving away from Earth in a random direction that would only come within reach of the Moon as a result of an unusually long-chance coincidence?

Dodge, however, has a better notion. Imagine yourself in a spaceship resting on Earth's equator. Earth is rotating on its

axis so that each point on the equator, including the spaceship, is moving about the axis at a speed of just about 0.46
kilometers per second. This is a supersonic speed (about 1.5
Mach) and if you were trying to hold on to an ordinary object
that was whirling you about at that speed, you would not be
able to hold on for the barest fraction of a second.

However, the Earth is very large, and the change in direction from the straight line in the time of one second is so
small that the acceleration inward is quite moderate. The
force of gravity upon the ship is strong enough to hold it to
Earth's surface despite the speed with which it is whirled. (It
would have to be whirled about the Earth at seventeen times
the speed before gravity would cease being strong enough to
hold it.)

But suppose the spaceship has a gravity screen plastered all
over its hull, and at a particular moment the screen is acti-
vated. Now, with no gravitation to hold it down, it is cast off
from the Earth like a clod of mud from a spinning flywheel.
It would move in a straight line tangent to the curve of the
Earth. The Earth's surface would drop below it, very slowly
at first, but faster and faster, and if you were careful to
activate the screen at just the right time, the ship's flight
would eventually intersect the Moon's surface.

Dodge does not mention that the Earth's curved motion
about the Sun would introduce a second factor, and that the
Sun's motion among the stars would add a third component.
These would represent comparatively minor adjustments,
however.

The landing on the Moon would be better than in the
previous plans, for a spaceship unaffected by the Moon's
gravity need not approach it at anything like its escape velocity. Once the ship is almost touching the Moon, the gravity
screen can be turned off and the ship, suddenly subject to the

Moon's relatively weak gravity, can drop a few feet, or inches, with a slight jar.

How about the return, though? The Moon rotates on its axis very slowly, and a point on its equator travels at 1/100 the speed of a point on Earth's equator. Using the gravity screen on the Moon will lend the spaceship only 1/100 the velocity it had on leaving Earth, so that it will take 100 times as long to travel from Moon to Earth as from Earth to Moon.

However, we can dismiss the whole notion. Albert Einstein promulgated his general theory of relativity thirteen years after Dodge's article was written, so Dodge can't be faulted for not knowing that a gravity screen is simply impossible. Scratch the Levitation Plan.

Dodge is most hopeful about his fifth scheme, "the Repulsion Plan." Here it is not just a matter of neutralizing gravity he hopes for, but some form of repulsive force that would actively overbalance gravitational attraction.

After all, there are two kinds of electrical charge and two kinds of magnetic pole, and in either case, like charges, or like poles, repel each other. Might there not be a gravitational repulsion as well as a gravitational attraction, and might not spaceships someday use a combination of the two, sometimes pushing away from an astronomical body and sometimes pulling toward it, and might this not help take us to the Moon?

Dodge does not actually say there might be such a thing as gravitational repulsion, and his caution is good, for, from the later Einsteinian view, it would seem that gravitational repulsion is impossible.

Dodge does mention light pressure, though, pointing out that it can, in some cases, counteract the force of gravity. He uses comets' tails as an example. Gravity would be expected to pull the tails toward the Sun but the solar light pressure pushes them in the opposite direction, overcoming gravitation.

Actually, he is wrong here, for solar light pressure, it turns out, is too weak to do the job. It is the solar wind that does it.

Light pressure might be used as a motive force, to be sure, but it would be too weak to work against the nearby pull of a sizable body or, for that matter, against air resistance. A spaceship would have to be in fairly deep space to begin with, and it would have to have sails that were extremely thin and many square kilometers in area.

As a way of lifting a spaceship from the Earth's surface toward the Moon, light pressure, or anything like it, is hopeless. Scratch the Repulsion Plan.

And that's all. Dodge is an intelligent, knowledgeable man who clearly understands science (as of 1903); yet if we consider only his five plans as he describes them, *not one* has the faintest hope of ever allowing human beings to travel from Earth to Moon.

Yet it has been done! My father was alive when that article was written and he lived to see human beings stand on the Moon.

How is that possible?

Well, have you noticed the word that Dodge omitted? Have you noticed that *he didn't see the truck?* He did not mention the rocket!

There was no reason for him to omit it. Rockets had been known for eight centuries. They had been used in peace and in war. Newton, in 1687, had thoroughly explained the rocket principle. Even earlier, in 1656, Cyrano de Bergerac in his story "A Voyage to the Moon" listed seven ways of reaching the Moon, and he *did* include rocketry as one of the methods.

How, then, did Dodge come to leave it out? Not because he wasn't sharp. In fact at the tail end of his article he was sharp-sighted enough to see something in 1903 that I have been laboring madly to get people to see now, eight decades later. (I'll discuss that in the next chapter.)

No, he didn't mention the rocket because the best of us don't see the truck sometimes. (I wonder, for instance, what trucks all of us are missing right now.)

Dodge *almost* got it with his recoil plan, but for the fact that he made an odd blooper. He thought that in order to get a decent recoil, the gun must fire a bullet that had a mass at least equal to itself, and that is wrong.

What counts in shot and recoil, action and reaction, is momentum. When a bullet leaves a gun with a certain momentum, the gun must gain an equal momentum in the opposite direction, and momentum is equal to mass *multiplied by velocity*. In other words, a small mass would produce sufficient recoil if it moved with sufficient velocity.

In rockets, the hot vapors that are ejected move downward with great velocity and do so continuously, so that the body of the rocket moves up with surprising acceleration considering the small mass of the ejected vapor. It still takes a large mass to begin with to deliver a comparatively small object to the Moon, but the disparity is far, far less than Dodge had feared.

Furthermore, the recoil effect is continuous for as long as the fuel is being burned and the vapors ejected, and that is equivalent to a projectile being moved along a cannon tube for hundreds of miles. The acceleration becomes small enough to be borne.

The possession of a reserve supply of fuel once the rocket is well on its way to the Moon means the rocket can be maneuvered; its descent to the Moon can be braked; it can take off for Earth again at will; and it can maneuver properly for entry into Earth's atmosphere.

And that is all, really, except for two coincidences, one

mild and one wild—and you know how I love to find coincidences.

The mild coincidence is this: In the very year that this article was written for *Munsey's Magazine*, Konstantin Tsiolkovsky began a series of articles in a Russian aviation magazine that went into the theory of rocketry, as applied to space travel specifically. It was the very first scientific study of the sort, so that modern astronautic rocketry had its start just at the time that Dodge was speculating about everything *but* rocketry.

The wild coincidence is this: Immediately after Dodge's article, in which he fails to mention the word "rocket" or to realize that it was the rocket, and the rocket alone, that would afford human beings the great victory of reaching the Moon, there is, of course, another article, and what do you think the title of that article is?

Don't bother guessing. I'll tell you.

It is "Rocket's Great Victory."

No, it isn't somebody else correcting Dodge's omission. It is a piece of fiction with the subtitle "The stratagem by which Willie Fetherston won a race and a bride."

"Rocket," in this story, is the name of a horse.

14

WHERE ALL THE SKY IS SUNSHINE

A friend of mine who is a publisher (almost all my friends seem to be writers, editors, or publishers, oddly enough—or perhaps that is not so odd) asked me to do a book of limericks for children.

"Clean ones," he said severely, having heard of some of my previous exploits in this direction. "That is, if you know how to do that kind."

"Of course I know how," I said, in the aggrieved tone I use when someone suggests there is some kind of writing I can't do if I put my mind to it.

"All right, then. I want fifty."

So a week later I brought in the limericks and he said, "Are you sure you've got fifty there?"

I couldn't believe it. He had actually presented me with the straight line I had dreamed of. Masking that, however, I said, in an offhand manner, "May I read you the last limerick?" and then did so:

TECHNOLOGY

50. Finish

Some say that my rhyme-schemes are shifty.
Some say that my meter is nifty.
I don't care either way,
For what I have to say
Is I'm finished. Please count them. There's fifty!

He was staggered. "I don't believe you," he said. "You're improvising. Let's see that."

I showed him the page. It was there.

"How did you know I would question the number?" he demanded.

"You ask? You with your unlovely, suspicious nature?"

(Not so, please understand, for he is a delightful person—as are just about all my writer/editor/publisher friends, to my unending happiness.)

And the best part of the situation is that my friend was so overawed by the aptness of the last limerick that he accepted all fifty without asking for a single revision or substitution.

This shows the power of a strong finish, and that brings me back to the 1903 article in *Munsey's Magazine*, which I was discussing in the previous chapter.

As you may recall, the article by Ernest Green Dodge, A.M., entitled "Can Men Visit the Moon?", listed five possible ways by which the lunar visit might be possible, every one of which was totally hopeless, but omitted the one method—rocketry—which was indeed possible, and which was eventually used.

In the last section of the article, however, he considers briefly the question: "What is the moon good for, even if man succeeds in reaching it?"

He points out that it is lifeless, airless, and waterless, and that it is "unspeakably cold" during its long night. In this he is perfectly correct, but he then goes on to make the curious error of saying that the Moon's temperature is "below the freezing point even at noon."

To be sure, it was not for another quarter of a century that direct and delicate measurements were made of the temperature of the Moon's surface. Still, considering that the Sun's rays strike the Moon's surface in as concentrated a manner as they strike the Earth, and that on the Moon there are no currents of air or water to carry off the heat and disperse it more or less evenly over the globe, and that on the Moon the sunshine continues without a break for fourteen days at a time in any one place, it was reasonable (and, in fact, inevitable), even in 1903, to conclude that high temperatures were reached during the lunar day.

Actually, the temperature of the Moon at its equator at noon is just a trifle above the boiling point of water.

Still, if Dodge was wrong in the letter, he was correct in the spirit, since a temperature that high would make the Moon even more unpleasant than one that was below freezing.

Dodge points out that despite all this, "men could abide there for a time in thick-walled, air-tight houses, and could walk out of doors in air-tight divers' suits." We call them space suits now, and the astronauts would live underground, in all likelihood, rather than in "houses," but the question still arises: Why go to all the trouble?

Dodge gives five answers, which manage to touch all the bases, in my opinion. Let's consider each of them in turn.

1. "Scientists would find in the lunar wastes a fresh field for exploration."

The year 1903 was, of course, in the heyday of polar exploration. Intrepid men were heading for both the North and South poles with concentrated determination. Robert E.

Peary was to reach the former in 1909 and Roald Amundsen the latter in 1911.

It may be that Dodge had that sort of exploration in mind and, if so, he didn't sufficiently count on technological advance. Satellites placed in orbit about the Moon sent back tens of thousands of photographs to Earth and from these a complete map of the Moon was worked out without any human being's leaving Earth. That left virtually nothing for the explorer of the classic Peary/Amundsen type to do.

The statement remains correct, however. Scientists *would* find the Moon a field for exploration, if we're talking about the search for subtle bits of geological, physical, and chemical evidence that would shed light on the past history of the Moon (and, for that matter, of the Sun, the Earth, and the solar system generally). This is already being done with the Moon rocks brought back by the Apollo astronauts, but it could be done far more effectively and in far greater detail if there were a permanent base on the Moon.

2. "Astronomers could plant their telescopes there, free from their most serious hindrance, the earth's atmosphere."

No argument here at all. We are planning to put moderately sizable telescopes in orbit about Earth, so one might suppose the Moon is not really needed. Suppose, though, we wanted to make use of a really large system of radio telescopy outside the growing interference of the radio waves emitted by Earth's heightening technology? The platform afforded by the far side of the Moon, with over three thousand kilometers of rock shielding it from Earth, would be unparalleled. (At the time Dodge was writing, Guglielmo Marconi's feat of sending radio waves across the Atlantic was less than two years old. We can't fault Dodge for not dreaming of anything like radio astronomy. Who could have done so, then?)

3. "Tourists of the wealthy and adventurous class would

not fail to visit the satellite, and costly hotels must be maintained for their accommodation.''

This was 1903, remember, when scions of the British ruling families were expected to go out to Africa or India and help build the Empire, and where the upper classes, bereft of honest work, were forced to engage in such trivialities as mountain-climbing and big-game hunting. (Do my prejudices show?) Nevertheless, I am sure there will be lunar tourism eventually, but I hope it will be, to as great an extent as possible, for all ''classes.''

4. ''Then it is quite probable that veins of precious metals, beds of diamonds, and an abundance of sulphur might be discovered on a world of so highly volcanic a character.''

Undoubtedly Dodge felt the Moon to be ''highly volcanic'' in nature because he assumed the craters to be the product of once active volcanoes. It is quite settled now that the craters are the result of meteoric bombardment in the early stages of the formation of the solar system, when bits of matter were still in the process of coalescing into worlds. Still, the larger strikes might well have broken the crust and allowed an upwelling of magma to form the lunar maria, so we'll let that go.

But ''veins of precious metals, beds of diamonds''? We now know that the Moon is no bonanza of silver, gold, platinum, or diamonds, but let us grant that Dodge could not have known that in 1903.

Even so, suppose that precious metals and diamonds *were* found on the Moon in great abundance. So what? The task of going out there to get them and then bringing them back would so add to their cost that it would be far cheaper to continue to scrabble it all out of Earth's weary crust.

Even if, somehow, technological advance made it possible to bring back all those ''precious'' things cheaply, it would be of no use. Dodge makes the mistake of confusing costly objects with valuable ones. Gold, silver, platinum, and diamonds are expensive and coveted only because they are rare. Diamonds can be used in industry as abrasives, platinum for laboratory ware, gold for fillings in teeth, and silver for photographic film, but if all these materials were as common as iron, all the uses we could think up for them would be insufficient to consume more than a small fraction of what was available.

There would be left their use as ornaments, for these things, gold and diamonds in particular, are undeniably beautiful. If, however, they were so common that there would be enough to supply everyone with such ornaments, they would no longer be desired. I don't think it is necessary to argue the point.

It follows, then, that it doesn't matter whether precious metals or gems are to be found on the Moon or not. What we *do* need to make it worthwhile to go to the Moon is some product that is *valuable*, rather than costly, and that can be used on the Moon or in nearby space.

Dodge is closer to the mark with his mention of sulfur. Sulfur is not a beautiful substance, or anything to be coveted for itself. It is, however, the basis of sulfuric acid which, setting aside the basic staples of energy, air, water, and salt, is the single most useful substance in the chemical industries.

But if Dodge is wrong in his examples, he is right in spirit, for the lunar crust can be used as a source of various structural metals, of clay, of soil, of cement, of glass, of oxygen—all of which can be prime building materials for structures in space. Indeed, if we are to have a space technology it will be supported in the main by mining stations on the Moon.

5. Next Dodge goes on to his final point, and we get the strong finish I spoke of in the introduction to this chapter. He says, "The world's population is capable of great increase . . .

And the world's need for motive power [energy] is increasing much more rapidly than the population.''

In this Dodge is utterly correct, but it is so obvious that any thinking person would have seen this, even in 1903, if he had bothered to think in that direction. Nevertheless I suppose that very few people, in 1903, would have felt any alarm concerning the matter. Western humanity was still riding the tide of nineteenth-century optimism, and it took World War I—still eleven years in the future—to shatter that.

Dodge, however, went on to put his finger on something that marks him as a man two generations ahead of his time. He says, "Our supply of coal and timber is limited, and will all too soon be exhausted.''

In Dodge's time, petroleum fractions were already being used as fuel, but still to only a very minor extent. Dodge did not foresee that the proliferation of the internal-combustion engine in everything from cars to planes would, within half a century, lift petroleum to the status of humanity's prime fuel, casting his "coal and timber'' into the shade.

That does not affect the cogency of his remark, however, for petroleum is far more limited in quantity than coal is, and, unlike timber, is not renewable. In short, coal and petroleum will alike some day be exhausted—petroleum long before coal—and timber alone cannot support our present population and technology. What, then, can we do?

Dodge is aware of alternate sources of energy. He says, "Waterfalls can do much. Windmills can do not a little.'' The implication is clear that they cannot, by themselves, do enough, however. There are other alternative sources that he doesn't mention: energy from waves, ocean currents, tides, temperature differences between the surface and depths of both land and sea, and so on. All are, or can be, useful, but perhaps all together will not be enough.

He does not mention (or even dream of, I suppose) nuclear

energy, though its existence had been discovered a few years earlier and H. G. Wells had speculated on the matter as early as 1901. Still, 1903 is a bit soon and I won't scold Dodge for missing it.

Dodge, however, goes on to say: "Solar engines, with concave mirrors to gather the sun's rays, have lately been put to practical use, and these in the future will accomplish wonders, yet even their resources, in our heavy, cloudy atmosphere are not boundless. But solar engines would work to much better advantage on the moon than on the earth."

I find this remarkable. He was forecasting solar power stations in space forty years before I was speculating about them in my short story "Reason" and sixty years before scientists began thinking seriously of them. I wonder if this might not be the very first reasonable mention of such a thing. (If any of my Gentle Readers knows of an earlier case, in science or in science fiction, I would like to hear of it.)

To be sure, Dodge thinks of solar power in terms of a concentration of the Sun's rays, a concentration that would deliver more heat to a small spot than would otherwise take place. Large mirrors would focus light upon a water reservoir, bringing it to a boil and producing steam. We would thus have a steam engine, with sunlight taking the place of burning coal as the producer of the steam. This would have the advantage that the Sun, unlike coal, would never be exhausted—or at least not for billions of years.

Dodge, in explaining the advantage of the Moon, clearly shows he is thinking of a steam engine, for he says the lunar engine would work better "owing partly to the absence of cloud and haze, but chiefly to the low temperature at which the condensed vapors could be discharged from the cylinders." (Here again, you see, Dodge labors under the feeling that the lunar surface would be very cold even with the Sun blazing down.)

WHERE ALL THE SKY IS SUNSHINE

By 1903, however, an increasingly important use for steam engines was that of turning a generator to produce an electric current. That use has become steadily more important in the decades since, but even in 1903 it might have been possible to wonder if sunlight on the Moon might not be turned into electricity directly, instead of having to move around Robin Hood's barn by way of the steam engine. After all, such direct conversion ("photoelectricity") was already known.

In 1840 the French physicist Alexandre Edmond Becquerel showed that light could produce certain chemical changes which, in turn, could produce electric currents. This wasn't quite a *direct* conversion of light to electricity but it showed a connection.

Something more direct involved the element selenium, which, along with its sister element tellurium, much resembles sulfur in its chemical properties. Of the two, tellurium, although the less common, was the first discovered.

Tellurium was discovered in 1783 by an Austrian mineralogist, Franz Joseph Müller. The discovery was confirmed in 1798 by the German chemist Martin Heinrich Klaproth, who was careful to give Müller full credit. It was Klaproth who gave the new element its name, tellurium, from the Latin word for Earth. He chose this name, apparently, because he had earlier discovered an element he had named uranium after the then newly discovered planet Uranus, which had in turn been named for the Greek god of the sky. The two elements were thus named for Earth and sky.

In 1817 the Swedish chemist Jöns Jakob Berzelius discovered a trace of a foreign substance in sulfuric acid, something he took to be a compound of tellurium. On closer examination he decided, in 1818, that what he had found was a substance containing not tellurium but an unknown element similar to it in properties. He wanted to balance the "Earth" that tellurium presented, and since "sky" had already been used, he chose "Moon" and named the new element selenium, from the Greek goddess of the Moon.

Selenium exists in different forms, depending on the arrangement of its atoms. One of these forms is silvery gray in color and is sometimes called "gray selenium." This shows certain metallic properties and it has, for instance, a slight tendency to conduct an electric current, though other forms of the element do not.

The tendency is quite small, but in 1873 Willougby Smith noted that when gray selenium is exposed to sunlight, the electrical conductivity of the element increases markedly. Once in the dark, the conductivity fades off, after a short interval, to the original low level again. The discovery aroused no particular interest at the time, but it was the first demonstration of a *direct* conversion of light to electricity.

Then, in 1888, the German physicist Heinrich Rudolf Hertz was experimenting with electric currents being forced to leap across an air gap (experiments that resulted in the discovery of radio waves). He found that when ultraviolet light shone on the negatively charged side of the gap, the electric current leaped the gap more easily than otherwise. This time the world of science listened, and Hertz is usually credited with the discovery of the photoelectric effect, though it existed in Smith's discovery of selenium's behavior fifteen years before.

The photoelectric effect comes about because light can knock electrons out of atoms, given the proper wavelengths and the proper atoms. Physicists had no explanation for the exact details of the effect until 1905, when Albert Einstein applied the then new quantum theory to the problem and won a Nobel Prize as a result.

However, the practical application of an observed phenomenon does not have to wait for a proper scientific explanation. In 1889, for instance, only a year after Hertz's demonstration of the photoelectric effect, two German physicists, Johann P. L. J. Elster and Hans Friedrich Geitel, were working together on the phenomenon.

They were able to demonstrate that some metals displayed the photoelectric effect more easily than others. (That is, electrons were more easily knocked free from some types of atoms than from others.) The alkali metals were most sensitive to the effect, and the most common alkali metals were sodium and potassium. Elster and Geitel therefore worked with an alloy of sodium and potassium and found that a current could be forced through it and across a gap without difficulty in the presence of visible light, but not in darkness.

This was the first "photoelectric cell" or "photocell," and it could be used to measure the intensity of light. The greater the intensity, the greater the electric current, and while the former was hard to measure directly, the latter was very easy to measure.

Though scientists could, and did, use the sodium-potassium photocell for scientific purposes, it wasn't very practical for everyday life. Sodium and potassium are highly active and dangerous substances and require the most careful handling.

At about the same time as Elster and Geitel were producing their photocell, however, an American inventor, Charles Fritts, was making use of the odd property of gray selenium which Smith had earlier observed. Fritts prepared little wafers of selenium, coated with a thin layer of gold. He incorporated them into an electric circuit in such a way that a current would flow only when the selenium wafers (another type of photocell) were illuminated.

Photocells, then, had been in existence for about four years when Dodge wrote his article in *Munsey's Magazine*. They were as yet out-of-the-way items, and I could scarcely blame Dodge if he had not heard of them. What's more, even if he had heard of them they scarcely seemed, at the time, to be more than small gadgets condemned to minor uses forever, and certainly not candidates for the large-scale conversion of sunlight to useful energy. In fact, despite magniloquent claims by Fritts, the selenium photocell converted less than one percent of the light that fell upon it into electricity—a fearfully low efficiency.

Still, the selenium photocells could be used for interesting minor purposes. The most familiar of these, to the general public, is the "electric eye."

Suppose that a door is equipped with some arrangement that can keep it open if allowed to work unimpeded. Suppose, further, that a small electric current can trigger a relay which will activate a larger one that will serve to pull the door closed against that arrangement. The small electric current runs through a circuit that incorporates a selenium photocell.

Next, suppose a small source of light at one side of the door sends a thin beam across the door to the selenium photocell on the other side. As long as that thin beam exists, the selenium photocell permits the passage of the small current that triggers the relay and keeps the doors closed.

If, at any time, there is an interruption of the light beam, even for a short time, the selenium photocell, momentarily in darkness, refuses to permit the current through. The small current fails, the relay is untriggered, and there is nothing to keep the doors closed. The door therefore moves open until the light is unimpeded and then it moves shut again.

A person approaching the door blocks the light with his body just before he reaches it. The door opens "of itself" in time to let him through and then closes again.

(I've always thought that if someone didn't know about electric eyes, you might have him watch you and, as you approach the door, you might shout "Open Sesame!" For a bewildered moment the observer might think he was in Ali Baba's tale in the *Arabian Nights*. That's what Arthur C.

Clarke means when he says that advanced technology is equivalent to magic to the uninitiated.)

The selenium photocell, and photocells generally, were not perceptibly improved until the mid-twentieth century. In 1948 scientists at Bell Telephone invented the transistor (see "Silicon Life After All" in *X Stands for Unknown*, Doubleday, 1984), and that changed everything. The transistor works because electrons can be knocked loose from atoms such as those of silicon or germanium. Research into transistors therefore meant research into something that might display a photoelectric effect.

This was not immediately obvious, to be sure, and when Darryl Chapin of Bell Telephone was looking for some source of power that could be used for telephone systems in isolated areas (something that could keep the systems working when conventional sources failed), he tried selenium photocells. It didn't work. Not enough sunlight could be converted to electricity to make it a practical hope.

Elsewhere in Bell Telephone, however, Calvin Fuller was working with the kind of silicon wafers used in transistors and it was found, more or less by accident, that sunlight would produce an electric current in them. Fuller and Chapin got together and produced the first practical solar cell.

Even the first solar cells they put together were 4 percent efficient, and eventually they got the efficiency up to 16 percent.

From that point on it became possible to dream of solar power in a manner that was more sophisticated than the concave mirrors that Dodge imagined. Suppose there were square miles of solar cells set up in some desert area, where sunshine was relatively steady. Would they not produce a steady flow of nonpolluting, neverending electricity in large quantities?

The catch is that a solar cell, though individually inexpensive, would, in the enormous quantities needed to coat a sufficiently large area of Earth's space, be prohibitively costly. Add to that the vast expense of proper maintenance after installation.

Nevertheless, solar cells have been used in minor ways such as the powering of orbiting satellites and have proved completely successful there. (I use a pocket calculator powered by solar cells, so that it has no battery and will never need one.)

What we must do is to make solar cells cheaper, more efficient, and more reliable. Instead of having to use single large crystals of silicon from which thin slices can be shaved off, it might become possible to use amorphous silicon made up of tiny crystals jammed together every which way—which would be *much* cheaper to produce.

And instead of setting up field upon field of solar cells, covering vast tracts of desert land where the air is not perfectly transparent (especially when the Sun is low in the sky), we might set them up on the Moon, where for two weeks at a time all the sky is sunshine and there is no air to interfere—or even in space, where there is hardly ever any night and where all the sky is sunshine nearly always.

In this way, Dodge's romantic dream might finally come true.

15

UP WE GO

I was in Boston the other week in order to help dedicate a new building at Boston University Medical Center. I am, after all, a Professor of Biochemistry there, and I ought to do *something* to prove it once in a while.

I was giving a luncheon talk, and before lunch I was interviewed and was told that the interview would appear the next day in *USA Today*— an issue I didn't happen to get. (Despite the reputation I have as possessing a monstrous ego, I usually manage not to see myself in the papers or on TV. I wonder why. Can it be that I *don't* have a monstrous ego?)

Someone said to me a couple of days later, "There was an interview with you in *USA Today* yesterday."

"Really?" I said. "I didn't see it. Was it interesting?"

"They said you don't fly," was the answer.

Big news! Everytime I am interviewed that item is featured. Not one interviewer in an uncounted number of years has failed to ask why I don't fly. The answer is, of course, that I have a fear of flying and have no interest in correcting that fear. But so what? What makes it such headline news every time?

When I suggest that my non-flying is unimportant, the interviewer is bound to muse over the curious fact that I flit from one end of the universe to the other in my imagination, yet I don't fly in real life.

Again, so what? I also write mysteries yet have never murdered anyone, and write fantasies without ever casting a spell in real life.

It gets a little wearisome to be a constant source of astonishment to everyone simply because I don't fly, and I sometimes think rather longingly that I would have been saved the trouble if the whole business of flying hadn't been thought up in the first place.

So let's consider the origins of this business of climbing into the sky and ask a question. What was the name of the first aeronaut?—No, the answer is not Orville Wright.

People have always wanted to fly. I suppose that what gave them the idea of doing so in the first place was that some living creatures *do* fly. There are three groups of animals now in existence that have developed true flight: insects, birds, and bats. (There was also a fourth group—the now extinct flying reptiles of the Mesozoic—but their existence was unknown to human beings till the nineteenth century.)

All flying organisms have one thing in common: wings that beat against the air. Each variety, however, has wings of a characteristic type. Human beings have supplied each wing type with mythical characters to suit, and have in this way managed to make very clear the relative popularity of the three. Thus: demons and dragons sport bat wings; fairies have gauzy butterfly wings; angels are equipped with large bird wings.

When human beings dreamed of flight, it might be by

sheer magic—carpets that flew at the sound of a magic word, wooden horses that flew at the turn of a magic peg, and so on. When a certain realism was demanded, the flying creature was imagined to have wings. The most famous example is Pegasus, the winged horse.

No one among the ancients seems to have noticed that all flying organisms were small. Insects are tiny, bats are usually mouse-sized, and even the largest flying birds are much smaller than many nonflying animals (or even nonflying birds such as ostriches). If this had been noticed, people might have deduced that there was no reasonable way in which really massive creatures could fly. There could be no winged pythons (dragons), no winged horses—and no winged men.

If people ignored this obvious (in hindsight) deduction, it was perhaps because it seemed to them that it was something other than smallness that was the key to flight. It was the birds that were flying creatures *par excellence*, and what birds had that no nonbird had was—feathers.

What's more, feathers are easy to associate with flight. They are so light that they have become a byword for that quality. "Light as a feather" is the cliché. A small fluffy feather will drift in the air, moving upward at every puff of wind almost as though it were trying to fly all by itself, even without the compulsion of inner life.

It seemed natural to suppose that if a man were going to fly, it was not so much wings with which he must supply himself as feathers.

Thus when Daedalus, in the Greek myth, wanted to escape from Crete, he manufactured wings by gluing feathers together with wax. He and his son Icarus, equipped with these wing-shaped conglomerations of feathers, were able to fly not through anything even vaguely aerodynamic, but through the aeronautical properties of the feathers. When Icarus flew too high and, therefore, too near the Sun, the wax melted, the feathers flew apart, and he fell to his death.

In actual fact, nothing but a bird has ever flown with feathers, nor has any human being or any human artifact ever flown by flapping wings, whether those wings were equipped with feathers or not. Active propulsion through the air, when it came, was by whirling propellers or by jet exhaust, methods no naturally flying organism uses.

It is not, however, necessary to fly in order to travel through the air and be an aeronaut. That is, it is not necessary to move independently of the wind. It suffices to move *with* the wind and to take advantage of updrafts to keep from descending under the inexorable pull of gravity—at least for a while. Such moving with the wind is "gliding."

Some birds which can fly perfectly well will, at times, glide for substantial periods of time, their wings stretched out and held steady. Anyone watching a bird do this might well get the impression that gliding is more fun than flying. Flying requires constant and energetic toil, while gliding is restful.

Some animals (the flying squirrel, the flying lemur, the flying phalanger and others) that cannot fly, can nevertheless glide. Their "flights" are, to be sure, quite limited when compared with those of the really successful fliers. Gliders are passive rather than active; they move under the control of the wind rather than of the will.

Nevertheless it is far simpler to emulate gliding than flying. Anything light and flat that presents a large surface to the air can be made to glide through the air. Make a gliding object light enough and large enough and devise a way of maneuvering it from the ground so as to have it take advantage of updrafts, and you have a kite—something that has been used as a toy in eastern Asia from ancient times.

The larger the kite, the greater its surface area compared to its total weight, the more extraneous weight it will carry. If a kite is made large enough (and yet strong enough) it can carry a human being. This is particularly so if the science of aerodynamics is developed and if a large kite (or ''glider'') is shaped in such a way as to increase its efficiency. In 1891 the German aeronaut Otto Lilienthal (1848–96) built the first glider capable of carrying a human being and sailed the air in it. (Five years later, alas, Lilienthal died in a glider crash.)

We all know that gliders look like flimsy airplanes without engines or propellers. In fact, in 1903, when the brothers, Wilbur Wright (1867–1912) and Orville Wright (1871–1948), invented the airplane, they did so by adding an engine and propellers to a glider which they had improved in various ways.

Can we say, then, that Otto Lilienthal was the first aeronaut? No, for if we do we would be wrong, for Lilienthal was not the first by over a century. It seems there is a third way of traveling usefully* through the air—not flying, not gliding, but *floating*.

From the dim beginnings of human thought, people must have noted that the smoke from a fire moves upward through the air and that, in the neighborhood of a fire, light objects—pieces of ash, bits of soot, fragments of feathers or leaves—move upward with the smoke.

Undoubtedly not one in a million of all those who observed this gave it any thought at all. The Greek philosophers did, however, since it was their business to make sense out of the universe. Aristotle, in his summary of the science of his day, worked it out this way about 340 B.C.:

There are five basic substances that make up the universe:

---

*Let's leave out of account falling or jumping off a cliff, being carried up by a tornado, or other such destructive events.

earth, water, air, fire, and aether. These are arranged in concentric shells. The earth is at the center, a solid sphere. Around it is a shell of water (not sufficient for a complete shell so that the continents are exposed). Around earth and water is a shell of air, and around that a shell of fire (ordinarily invisible, but occasionally seen as a flash of lightning). On the very outside is the aether, which makes up the heavenly bodies.

Each substance had its place, and when it was for some reason out of place, it hastened to return. Thus any solid object suspended in air fell to the earth as soon as it was released. On the other hand, water or air trapped underground will tend to rise if released. In particular, a fire once started will strive to reach its place up above the air. That is why flames leap upward. Smoke, which contains much in the way of fiery particles, also moves upward through the air and with such force that it can carry light nonfire particles with it, at least for a while.

This was a very sensible explanation, given the knowledge of the time, and the matter was questioned no more.

Of course, there were problems with the explanation. A stone released on the surface of the pond sinks through the water and comes to rest on the earthy bottom, as one would expect by Aristotle's theory. Wood, however, which, like stone, is solid and should therefore be thought to be a form of earth, would, if released on the surface of a pond, remain there, floating on the water indefinitely.

An Aristotelian explanation might be that wood contains an admixture of airy particles that imparts enough natural upward movement to make it float—and that's not such a bad attempt at explanation, either.

The Greek mathematician Archimedes (287–212 B.C.) however, worked out the principle of buoyancy. This explained floating by comparing densities of solid objects with water. A solid that was less dense than water would float in water. Floating was thus handled in quantitative terms, and not merely qualitatively. By measuring density one could not only predict that a substance would float, but also exactly how far it would sink into water before coming to a floating rest. It also explained why an object that did not float nevertheless was reduced in weight when immersed in water, and by just how much its weight would be reduced.

In short, Archimedes's explanation was much more satisfactory than Aristotle's was.

It followed, then, that the principle of buoyancy might be applied to air as well. Something less dense than air would rise in air, just as something less dense than water would rise if immersed in water. This analogy occurred to no one, however, for eighteen centuries after Archimedes's time, simply because no one thought of air as in any way analogous to water. Air was not recognized as a substance, in fact.

The turning point came in 1643, when the Italian physicist Evangelista Torricelli (1608–47) demonstrated that the atmosphere (and, therefore, any sample of air) had measurable weight. The atmosphere could, in fact, support a column of mercury thirty inches high (such a column being the first barometer). In this way, air was finally recognized as matter— very attenuated matter, but matter.

One could reason from Torricelli's discovery that if a given volume of any substance weighed less than that same volume of air, that substance would be less dense than air and would rise.

Then, in 1648, the French mathematician Blaise Pascal (1623–62) persuaded his brother-in-law to climb a local mountain with two barometers and showed that the weight of the atmosphere declined with height. In fact, the weight declined in such a way that it was plain that the density of air decreased with height.

This meant that a substance less dense than air would rise until it reached a height at which its density matched that of the thinning air. It would then rise no higher.

So far, so good, but there was no substance known that was less dense than air. Even the least dense of the ordinary liquids and solids known to human beings of the time was hundreds of times as dense as air.

But what about no substance at all? What about nothingness?

When Torricelli constructed his barometer, there was a space above the top of the mercury column that contained nothing but a trace of mercury vapor. It was the first vacuum ever created by human beings, and a vacuum is certainly less dense than air.

What's more, in 1650 a German physicist, Otto von Guericke (1602–86), invented an air pump which, for the first time, could (with a lot of hard work) produce a considerable volume of vacuum.

About 1670, therefore, an Italian physicist, Francesco de Lana (1631–87), became the first to suggest the construction of something that would float in air. He pointed out that if a thin copper sphere were to be evacuated, then the total weight of the copper averaged over the volume of the sphere (with no air inside to add to the weight) would be less than that of an equal volume of air. Such an evacuated sphere would rise. If the spheres were made large enough, and if enough of them were attached to some sort of light gondola, the whole would rise into the air carrying a man.

Actually, the scheme was not practical. If a copper sphere were thin enough to rise upon evacuation, the copper would be far too thin to withstand the air pressure to which it would be exposed. It would collapse as it was evacuated. If the sphere were thick enough to withstand the air pressure, it would be too thick to average out to less than air density under any practical circumstances. Nevertheless, De Lana was the first to envisage a "balloon" (an Italian word meaning "a large ball").

De Lana's notion of using a vacuum for buoyancy was not the end, however. In the 1620s the Flemish chemist Jan Baptista van Helmont was the first to recognize that there were different gases (he was the first to use that word) and that air was not unique. In particular he was the first to study the gas we now call carbon dioxide (see Chapter 9).

It might be that a gas existed that was less dense than air and that would therefore float in air, but if so, carbon dioxide was not it for it happens to be about 1.5 times as dense as air. It was not, however, till the 1760s that anyone measured the densities of particular gases, so it was not until then that anyone could reasonably speculate on gas-filled balloons.

In 1766 the English chemist Henry Cavendish (1731–1810) produced a gas by the action of acids on metals. He found it to be quite inflammable and he therefore called it "fire gas." He measured its density and found it to be only 0.07 times that of air. This set a record for the low density of normal substances under Earth surface conditions that has endured to the present day.

In 1784 Cavendish found that hydrogen, on burning, formed water, so the French chemist Antoine Laurent Lavoisier (1743–94) named it "hydrogen" (from Greek words meaning "water-producer").

Let us suppose now that we have a volume of air weighing 1 kilogram. That same volume of vacuum would weigh 0 kilograms, and if we can imagine ourselves hanging weights on that volume of vacuum, we could hang 1 kilogram on it to bring its weight up to that of the same volume of air (and thus bring the average density of the system up to that of air). This would just keep the vacuum from rising.

If, instead, we took the same volume of hydrogen, that would have a weight of 0.07 kilograms, and we would have to have a weight of 0.093 kilograms upon it to make its weight equal to that of the same volume of air and keep it from rising. Hydrogen, in other words, would have a surprising 93 percent of the buoyancy of vacuum—and it is a lot easier to fill a container with hydrogen than to evacuate it.

What's more, hydrogen, under ordinary conditions, would have the same number of molecules per unit volume that air has. Although hydrogen molecules are lighter than air molecules, hydrogen molecules move faster, and in the end the momentum of the molecules (and therefore the pressure) is the same in both cases.

This means that whereas vacuum, when used for its buoyancy effect, must be contained by metal thick enough to withstand air pressure, which adds prohibitive weight to the system, the condition is quite different for hydrogen. The pressure of hydrogen inside the container would just balance the pressure of air outside, so that the container itself could be as flimsy and light as possible—as long as it was reasonably gas-tight and didn't allow hydrogen to diffuse outward, or air to diffuse inward.

You would think, then, that as soon as Cavendish had discovered the low density of hydrogen, he, or possibly someone else, would have thought of its buoyancy effect and gone about the building of a balloon. Not so! However clear hindsight may be, foresight can be remarkably dull, even for a first-class scientist like Cavendish.

In point of fact, hydrogen ended up having nothing to do with the invention of the balloon.

This brings us back to the earlier question about smoke rising. Why does smoke rise when it is made up of particles that are individually denser than air, and contains gases, like carbon dioxide, that are also denser than air?

The key to the answer came in 1676 when a French physicist, Edmé Mariotte (1620–84), noted that air expands when it is heated. If a given quantity of air expands, its fixed amount of mass is spread out over a larger volume, which is another way of saying that its density decreases. In other words, warm air is less dense than cold air, and has a buoyancy effect. The warmer the air, the greater the buoyancy effect. This was made plainer in 1699 by the studies on gases of a French physicist, Guillaume Amontons (1663–1705).

An ordinary wood fire heats the air about it to a temperature of up to 700 C., and the density of air at such a temperature is only half that of ordinary air. Such hot air has about half the buoyancy effect of hydrogen (or vacuum, for that matter). The column of hot air rises vigorously and carries with it other gases and the light materials that make up smoke.

There are advantages of hot air over hydrogen that tend to make up for the fact that hot air is not quite as buoyant. Hot air is easily obtainable; all you need is a fire. Hydrogen, on the other hand, is comparatively difficult to collect in quantity. Furthermore, hot air is not inflammable, while hydrogen is actually explosive. On the other hand, the buoyancy of hydrogen is permanent, whereas hot air loses buoyancy rapidly as it cools, so that you must not merely have a fire at the start, but keep it going as long as you want to stay aloft.

One would suppose that as soon as the low density and, therefore, buoyancy of heated air was recognized, someone would conceive of a balloon and try to build one, but that's hindsight again. It took a century for the thought to occur to anyone.

The brothers Joseph Michel Montgolfier (1740–1810) and Jacques Étienne Montgolfier (1745–99) were two of the sixteen children of a well-to-do paper manufacturer. An ancestor of theirs (according to family tradition) had learned the technique of paper manufacture while in a prison in Damascus at the time of the Crusades, and had brought it back from the East.

The brothers had watched objects rising in the hot air produced by fires, and the older brother had been reading about the new findings of gases and, somehow, they got the notion of a hot-air balloon.

First they tried it at home. In November 1782 they burned paper under a silk bag with an opening at the bottom. The air within the bag heated up and it rose to the ceiling. They repeated the experiment in the open air and the bag rose to a height of 20 meters (that is, the height of a six-story building). They kept trying-larger and larger bags and finally decided on a public demonstration.

On June 5, 1783, in the market place of their home town, the brothers used a large linen bag, 10.5 meters (35 feet) in diameter, and filled it with hot air. They had invited everyone in town to witness the experiment and the crowd saw the balloon rise 2 kilometers (1.2 miles) into the air and stay in the air for ten minutes during which it slowly descended as its air content cooled. It traveled 2.5 kilometers (1.5 miles) during its descent. It was an electrifying demonstration and it created a sensation.

The news traveled to Paris, and there a French physicist, Jacques Alexandre César Charles (1746–1823), was told of it. Instantly he thought of hydrogen.

On August 27, 1783, he prepared for a demonstration of his own in Paris. He used 225 kilograms of acid and 450 kilograms of iron pellets to produce the hydrogen. The gas fizzed up madly and rose into the open neck of the bag held over it, displacing most of the air. When the balloon was released, it rose a kilometer into the air. The hydrogen slowly diffused out of the bag, but as it lost height it traveled 25 kilometers (15 miles) in 45 minutes before coming to earth.

When it did so, the peasants of the neighborhood, who had heard nothing of ballooning and could only assume a vehicle flying through air (a UFO, we would call it today) to be carrying invaders from some other world, bravely attacked it with scythes and pitchforks and destroyed it.

These balloons were simply bags. It was clear, though, that one could hang weights on the balloons, which would slow their rise and limit their height, yet not destroy the buoyancy effect altogether. The Montgolfiers, with this in mind, planned the most sensational demonstration yet for the French court at Versailles.

On September 19, 1783, they made use of a balloon of record size, one that was 13 meters (43 feet) in diameter. Under it was a wicker basket into which were placed a rooster, a duck, and a sheep. The wicker basket also contained a metal brazier within which there was fuel. The fuel was set on fire, and the balloon filled with hot air. It was released, and up it went before the eyes of a crowd of 300,000 people (including the King and Queen of France and Benjamin Franklin). The balloon, with its animal load, traveled for 3 kilometers (nearly 2 miles) before coming down once the fuel was consumed and its air content cooled. The first person on the scene when the balloon landed was a young French physicist, Jean François Pilâtre de Rozier (1756–85).

The animals were not harmed and they were the first living things ever to be carried through the air by a man-made contraption.

But if a sheep, why not a man? That was clearly the next step. King Louis XVI, who was fascinated by the demonstration, was nervous about manned flight. It seemed too dangerous, and he suggested that condemned criminals should be asked to volunteer for the flight with the promise of a pardon if they survived.

Pilâtre de Rozier, however, craved the honor. He and a French nobleman, François Laurent, Marquis d'Arlandes, argued their case with Queen Marie Antoinette, convinced her, and she convinced the king.

On November 20, 1783, Pilâtre de Rozier and the Marquis d'Arlandes got into a wicker basket and went up in a hot-air balloon. They were carried 8 kilometers (5 miles) in 23 minutes, and landed unharmed.

These two were the first aeronauts, 120 years before the Wright brothers and 108 years before Lilienthal.

Pilâtre de Rozier earned another remarkable first a year and a half later.

On January 7, 1785, the English Channel was crossed for the first time by balloon. On board were a Frenchman, Jean Pierre François Blanchard (1750–1809), who was the inventor of the parachute, and an American, John Jeffries (1744–1819).

On June 15, 1785, Pilâtre de Rozier and another Frenchman, Jules Romain, tried to duplicate the feat in the other direction. However, the fire used to heat the air in his balloon set the fabric of the balloon aflame, and the two aeronauts fell 1,500 meters to their death.

So the first aeronaut, like a real-life Icarus, died in the first aeronautical disaster.

PART VI
CHRONOLOGY

16

THE DIFFERENT YEARS OF TIME

One more story about my bypass operation and then I'll shut up about it. (Well, I *may*.)

When I found out that for a period of time I was going to be in a heart-lung machine, I worried about whether the anesthesiologist would take proper care to see that my brain, in particular, received an ample supply of oxygen. The brain consumes one quarter of the oxygen the body uses and it seemed to me that even a temporary, small shortage would damage it marginally.

I wanted *no* damage, not the most marginal. I've been making an enjoyable life out of the exceedingly sharp edge of my brain and I didn't want it blunted.

I expressed my fears to my family internist, good old Paul, the sweetest M.D. in the world.

"Don't worry, Isaac," he said. "I'll see to it that everyone understands the situation, and I will personally test you."

And so he did. I don't remember it, but he told me what

happened. Although I didn't really come to with full consciousness before ten the following morning, I did stir now and then at earlier times so there were momentary fits of responsiveness followed by retreat into my anesthetized semicoma.

At 10 P.M., some hours after the operation was completed, my eyes flickered open and Paul was standing there. "Hello, Paul," I whispered huskily (he said).

He leaned toward me, "Hello, Isaac. Make up a limerick about me."

I blinked a few times, then whispered:

"There was once an old doctor named Paul
With a penis exceedingly small—"

Whereupon Paul said austerely, "Go no further, Isaac. You pass."

When he told me the story the next morning, I was greatly relieved for it meant I could continue to write. And here goes—

You have probably encountered a number of times a little teaching device whereby you allow the history of the Earth to be compressed into a year and then mark off at what time of the year various landmark events in Earth's history took place. This gives you a more easily grasped view of the sweep of time and the relative chronological position of various bits and pieces of it.

Naturally you find that humanity came into being very late, on the last day of the year, and you get a dramatic notion of our insignificance as an item in the chronology of the planet.

This is not a peculiarity of Earth history alone, but of every facet of any kind of history. We always see things close to us in great detail, while as we look farther and farther away, we see matters more and more fuzzily and view it with less and less interest. Contemporary times always seem to be very long and detailed, while long past times seem short and uninteresting.

For instance, pick up a school history of the United States that deals with the period of time from Columbus's voyage in 1492 to the present. Divide the book at the Declaration of Independence and notice how many pages are given the exploratory and colonial periods, and how many pages are given the period of the United States as an independent nation. I haven't got such a book to check, but my guess is that it would divide up into a 1 to 6 ratio.

That makes sense for a number of reasons and I don't quarrel with it, but the average schoolchild (or adult, for that matter), leafing through such a book, couldn't help but get the vague notion that the strictly chronological division is similar: that the United States as an independent nation has endured for much longer than the relatively brief colonial period that preceded it.

To see what the situation really was like, let's use the trick of squeezing a period of time into an arbitrary year, and compressing the chronology, without distortion, into the days of that year.

Thus the first permanent settlement of Englishmen in what is now the territory of the United States was at Jamestown, Virginia, on May 14, 1607. Call that New Year's Minute—12:01 A.M., January 1. Call the present moment Old Year's Minute—11:59 P.M., December 31. The time lapse from the settlement of Jamestown till now (at this time of writing) is 377 years. That means each day of our "United States Year" is equal to 1.03 real years.

CHRONOLOGY

1. The United States Year

Jamestown settled	January 1
The landing at Plymouth Rock	January 13
The British take New Amsterdam	February 25
Philadelphia founded	March 2
Georgia (last of the 13 colonies) founded	May 3
French driven from North America	June 1
Declaration of Independence	June 14
British recognize American independence	June 21
Louisiana Purchase	July 10
Missouri Compromise	August 1
Gold discovered in California	August 22
Civil War begins	September 4
United States enters World War I	October 26
Stock market crash	November 9
Pearl Harbor	November 21
V-J day	November 25
Kennedy assassinated	December 12
Nixon resigns	December 23
United States conquers Grenada	December 31

If you study this table, it may surprise you that for nearly half the time that men of English descent have lived on what is now American territory, there was no United States. It was not till very nearly the middle of the year that the United States was legally independent by treaty with Great Britain.

The thing that *I* find most surprising, however, is that when V-J day came, it was not yet December. After all, I remember V-J day as though it were yesterday. How can a whole United States Month have passed since then? Well, it has. Almost thirty-nine years have passed since V-J day as I write, and that is almost a fifth of the total duration of American independence.

Almost, it makes me feel old.

We can do this sort of thing on longer and longer scales. Suppose, for instance, that we begin with Columbus's landing at San Salvador on October 12, 1492.

That would cover the entire stretch of time in which North America was penetrated, explored, and occupied by European powers. If we compress that into "the North America Year" we find that we cover a period equal to 492 years, so that each North America Day is 1.348 real years long.

2. *The North America Year*

Columbus lands at San Salvador	January 1
Ponce de Leon discovers Florida	January 16
Cortez takes Mexico	January 20
De Soto sights the Mississippi River	February 6
Drake sails along the California coast	March 5
Jamestown founded	March 26
Declaration of Independence	July 30
Civil War begins	September 27
Pearl Harbor	November 29
Kennedy assassinated	December 15
United States conquers Grenada	December 31

Notice that during the first third of all the time that Europeans were trudging along the coasts and into the interior of the North American continent, those Europeans were almost all Spanish. It wasn't till March 25 that Englishmen came to North America to stay.

And for 5/9 of the time that Europeans of any kind have been in North America, there was no United States.

There was, of course, history before the United States and even before European North America. Lots of it. Such history

247

is usually divided into ancient times, medieval times (or the Middle Ages), and modern times. Most people, I suspect, assume that these three periods are of approximately equal duration. If anything, they might imagine that modern times is the longest of the three, because it always bulks longest in the history books.

Let's see, then . . . History begins with writing. Writing makes possible the recording of chronicles, the giving of names, dates, places. Without writing, we must infer matters from artifacts, and we can never determine the kind of detail that makes history what it is.

As far as we know now, the first writing was invented by the Sumerians, possibly as early as 3,200 B.C. Let us therefore begin "the History Year" with 3,200 B.C. as January 1. That gives us a stretch of 5,184 years to the present, so that each History Day is equal to 14.2 real years.

3. The History Year

Writing invented	January 1
First pyramid built in Egypt	February 1
Sargon establishes first empire in Asia	March 2
Hammurabi's law code	April 16
Egyptian Empire at its height	May 7
Trojan War	May 21
David becomes King of Israel	June 4
Homer composes *Iliad*	June 18
Rome founded	June 21
Assyrian Empire at its height	June 26
Nebuchadnezzar destroys Solomon's Temple	July 4
Cyrus founds the Persian Empire	July 7
Battle of Marathon	July 10
Athens in its Golden Age	July 13
Alexander the Great conquers Persia	July 21

THE DIFFERENT YEARS OF TIME

Rome defeats Carthage; dominates Mediterranean	July 31
Julius Caesar assassinated	August 11
Jesus crucified	August 16
Roman Empire at maximum extent	August 22
Constantine founds Constantinople	September 6
End of west Roman Empire and of ancient times	September 16
Charlemagne crowned Emperor	October 9
William of Normandy conquers England	October 28
Crusades begin	October 30
Magna Carta	November 7
Black Death strikes Europe	November 17
Fall of Constantinople	November 24
Discovery of America; end of medieval times	November 27
Protestant Reformation begins	November 29
English "Glorious Revolution" 1688	December 9
Declaration of Independence	December 17
Fall of the Bastille	December 18
Human being lands on the Moon	December 31

Notice that by the time half the History Year is over, the great days of Greece have *not yet come*. We and the ancient Greeks alike are products of the second half of history.

The whole first half of the History Year is dominated by Asian kingdoms. Greece occupies the History Month of July, Rome that of August. The History Year is 70 percent over before "ancient times" comes to an end. In other words, ancient times (although it gets the least attention in modern books) lasts more than twice as long as medieval times and modern times together.

Whereas ancient times lasts 8.5 History Months, medieval times lasts just over 2 History Months, and modern times has endured just over 1 History Month.

In the History Year, I mention the English Revolution of

1688, the American Revolution of 1776, and the French Revolution of 1789. Each did its bit to establish the modern rule of liberalism and human rights. But notice that it is only in the last half of December of the History Years that this has been established and, even so, in only a small part of the world and there, precariously.

One can but sigh.

To be sure, civilization antedates writing. The word "civilization" comes from the Latin word for "citizen"—that is, "city-dweller." Let us then date civilization from the establishment of the first small cities (Jericho, for instance, in Palestine).

The beginning of civilization may be put (a bit arbitrarily) at about 8,000 B.C., or 10,000 B.P. ("Before the Present"). About then, groups of people in western Asia learned to domesticate plants and animals. They turned from food gathering to agriculture and herding. This made larger concentrations of people possible in a given area and led inevitably to the founding of cities.

If we start "the Civilization Year" at 8,000 B.C., this gives us a duration of 9,984 years and makes each Civilization Day 27.35 real years long.

4. The Civilization Year

Earliest cities founded	January 1
Earliest pottery we know of	February 6
Agriculture reaches southeastern Europe	March 14
Agriculture in the Nile Valley	April 20
Beginning of the use of metals	May 8
Traditional date of biblical creation	May 26
Invention of writing	June 25
Beginning of the Bronze Age	July 2

THE DIFFERENT YEARS OF TIME

The Great Pyramid built	July 20
Iron Age begins	September 12
Solomon builds the temple at Jerusalem	September 25
Jesus crucified	October 21
End of ancient times	November 6
End of medieval times	December 13
Declaration of Independence	December 23
Assassination of Kennedy	December 31, 6 A.M.

The traditional date of biblical creation referred to in the Civilization Year is 4,004 B.C., as determined by Archbishop Ussher and as is still given in most editions of the King James Bible. By that time, however, civilization had endured for two fifths of its total span.

More than half the period of civilization had passed before the Great Pyramid was constructed. We think of the stretch of time from our lifetime to the pyramids as enormous, but before that there stretched an even longer period of pyramid-less civilization. It was not only pyramid-less, but was totally illiterate.

To be sure, this first half of civilization, without writing or pyramids, was crude and rudimentary by our standards and existed only over a small patch of the world, but we must not sneer at it. We are what we are today because we built on the achievements of those illiterates. An impartial assessment of what they did might lead to the conclusion that it was they who had the harder task than we, and who accomplished more on the basis of what they had to work with.

In fact, even before there were cities and agriculture, human beings made great advances and, notably, showed that they were great artists and ingenious hunters and toolmakers. *Homo sapiens sapiens*— "modern man"—throughout his/her existence demonstrated great ingenuity and adaptability, and it is highly arbitrary to define civilization in terms of one particular advance such as city building. The history of "modern man," is one of steady advance.

What about "the Human Year," then? Suppose we start it at 35,000 B.C. (37,000 B.P.), by which time "modern man" was the only hominid living on Earth—though only on the continents of Africa and Eurasia. The total duration of 36,984 years means that each Human Day is 101.3 real years long.

5. The Human Year

Homo sapiens sapiens begins domination of Earth	January 1
Beginnings of representational art	April 9
Human beings migrate into Australia and America	May 28
Cave paintings at their height	August 10
Human beings complete settlement of Americas	September 14
Civilization begins	September 24
Great Pyramid built	November 17
End of ancient times	December 16
End of medieval times	December 26
Declaration of Independence	December 29
Pearl Harbor	December 31, 10 A.M.

More than half the stretch of the Human Year elapsed before the time of the great cave paintings came, and nearly three quarters of the stretch elapsed before what we call civilization began. Only the last quarter of the history of "modern man" showed any civilization anywhere.

The United States has existed for only two Human Days.

*   *   *

252

There were, of course, hominids before modern man. For that matter there was *Homo sapiens* before modern man. The so-called Neanderthal men *(Homo sapiens neanderthalensis)* were of the same species as we, and could (and presumably did) interbreed with our ancestors. Their genes must still rest among us.

And before the Neanderthals, there were other, smaller, lesser-brained species of the genus *Homo,* and before that there were creatures with still lesser brains who were not *Homo* but who were still hominids, and who walked upright, had hands like ours and were, in general, closer to us in anatomic detail than they were to the apes.

The first hominids we can be sure of were australopithecines, living in southern and eastern Africa, no larger than children of our own species but walking upright as we do and having their hands free to explore and manipulate the universe.

They may have made their appearance about 4,000,000 years ago and, while there is talk of still earlier hominids, I will start "the Hominid Year" at 4,000,000 B.P. This would mean that each Hominid Day would be 10,960 real years in length.

6. *The Hominid Year*

Australopithecines appear	January 1
Genus Homo *(Homo habilis)* appears	July 2
Homo erectus (Peking man) appears	August 15
Fire comes into use	November 15
Homo sapiens (Neanderthal man) appears	December 18
"Modern man" appears	December 26
"Modern man" only hominid on Earth	December 28
Civilization begins	December 31, 2 A.M.
History begins	December 31, noon

CHRONOLOGY

During the first half of the Hominid Year, australopithecines were the only hominids to exist. It was only after 95
percent of the Hominid Year had passed that *Homo sapiens*
made its appearance. "Modern man" is a creature of the last
week only and all of civilization is crowded into the last day.

Incidentally, for seven eighths of the time during which
hominids existed on Earth, they did so without the use of fire.
The development of that use was the greatest achievement of
pre-*sapiens* days. The achievement was that of *Homo erectus*,
for the remains of campfires have been found in the caves
that housed the bones of Peking man.

Hominids are not the only organisms that have left fossil
remains through which we can trace paleontological history.
Before the hominids were earlier primates and other mammals before them, and nonmammals and invertebrates. A rich
fossil record stretches back for about 600,000,000 years.

Let us then set up "the Fossil Year" and begin it at
600,000,000 B.P. Each Fossil Day would thus be 1,644,000
real years long.

7. The Fossil Year

Fossils appear, all invertebrate	January 1
First vertebrates appear	March 1
First land plants appear	April 12
First air-breathing fish appear	April 30
First forests appear	May 4
First land vertebrates (Amphibia) appear	May 12
First reptiles appear	July 1
First dinosaurs appear	August 30
First mammals appear	September 5
First birds appear	September 29
Flowering plants appear	October 30

THE DIFFERENT YEARS OF TIME

Extinction of the dinosaurs	November 21
Large mammals dominate Earth	November 28
First hominids appear	December 27
Fire comes into use	December 31, 4 P.M.
Neanderthal man appears	December 31, 10 P.M.
"Modern man" appears	December 31, 11:15
Civilization begins	December 31, 11:50

As you see, for the first quarter of the Fossil Year no land life existed, and there were no land vertebrates till three eighths of the Fossil Year had elapsed.

The reptiles appeared only when the Fossil Year was half over, and the dinosaurs dominated the Fossil fall. The hominids are creatures of the last four Fossil Days, modern man of the last 45 Fossil Minutes, and all of history is crowded into the last 10 Fossil Minutes.

But there was life before the fossils. The only reason fossils appeared so suddenly 600,000,000 years ago is that there was a preceding evolutionary efflorescence that produced shells and other hard parts of increasingly complex organisms, and it was these parts that fossilized easily.

Before those complex animals there were small, soft-bodied organisms that did not fossilize well, and before them microscopic organisms that could leave only the barest traces.

These traces have been found, however, and paleontologists have followed life far back to near the beginning of Earth's existence. We will therefore set up "the Earth Year" and start it 4,600,000,000 years ago, at which time Earth first assumed more or less its present form (as did the Sun, and the solar system generally). Each Earth Day is therefore 12,600,000 real years long.

CHRONOLOGY

8. *The Earth Year*

Earth assumes its present form	January 1
Primitive bacteria develop	April 1
Photosynthesis begins in blue-green algae	May 21
Multicellular organisms develop with simple cells	July 24
Cells with nuclei (eukaryotes) develop	October 11
True animals develop	October 27
Rich fossil record begins	November 12
Land life (plants) appears	November 26
First dinosaurs appear	December 14
Extinction of dinosaurs	December 26
First hominids appear	December 31, 4:30 P.M.

You see, then, that if we take the Earth as a whole, it passed perhaps a quarter of its lifetime as a lifeless globe. During nine tenths of its lifetime, the *land* remained lifeless. It is a testimony to the difficulty of dry land as a vehicle for life that land life is a product of only the last Earth Month.

The dinosaurs are creatures of Earth mid-December only, and the entire stretch of hominid existence is a matter of only the last 7½ Earth Hours. Modern man has been on Earth for only the last 5¾ Earth Minutes, and history is a matter of only the last 35 Earth Seconds or so.

Nor are we finished even yet. I'll carry on in the final chapter with still vaster sweeps of time.

17

THE DIFFERENT YEARS OF THE UNIVERSE

I've got to tell you about the only time during a warm forty-five-year friendship with my fellow writer Lester del Rey that I stopped him cold.

It's not easy. He never allows any verbal blow without an immediate counterblow; he's never at a loss for a retort, and he never hesitates to make it—except, in my case, that once.

He and I and two other friends were in a taxi and somehow I was on the subject of my patriarchal father and his countless moralistic admonitions to me, for he had firmly believed that only by endlessly subjecting me, in my impressionable childhood, to the teachings of the great Jewish sages, could he prevent me from losing my way in the thickets of immorality and vice.

"Remember, Isaac," he would say, in that melodious singsong with which Jewish moral lessons were inculcated, "that if you hang around with *bums*" (the word was always emphasized strongly so as to indicate the utmost contempt and moral revulsion) "you may think you will change them into decent people, but you will *not*. No! Never! Instead, if you hang around with *bums*, they will change *you* into a *bum*."

Whereupon Lester interposed instantly to say, "So why do you still hang around with bums, Isaac?"

And I replied without hesitation, "Because I love you, Lester, that's why."

That was the first and only time in my long experience with him that Lester burst out laughing so hard he found himself unable to answer. What's more, I had the two other guys in the cab (also laughing, of course) as witnesses.

I thought of that the other day when I was being interviewed by someone who said to me, "Of all the different kinds of writing you do, Dr. Asimov, what sort do you enjoy the most?"

I have been asked that many times before (I have been asked *everything* by interviewers many times before) so I didn't have to do any thinking. I said, "I most enjoy my monthly essays for *Fantasy and Science Fiction* magazine. I've been doing them for over a quarter of a century without missing a deadline."

The interviewer looked doubtful. "Is that because they pay well?"

"No," I said. "As a matter of fact, the word rate is lower for those essays than for anything else I write, but I'd do them for nothing, if I had to."

"But why?"

And the answer came again without hesitation. "Because I love them, sir, that's why."

And I do. It may be that there is a Gentle Reader somewhere who secretly believes that no one in the world enjoys these essays as much as he (or she) does. If so that Gentle Reader is wrong. *I* enjoy them more.

THE DIFFERENT YEARS OF THE UNIVERSE

With that, we'll continue from where we left off in the preceding chapter.

In the preceding chapter I took various significant lengths of time—American history, the history of civilization, the history of hominids, and so on—compressed each into a year and marked off significant events (without relative distortion) along the length of the year. It gave, more accurately than the ordinary way of dealing with dates, what I thought was a dramatic notion of what happened.

The last compression I undertook was the 4.6-billion-year history of the planet, Earth, made into one "Earth Year." It showed, for instance, that if Earth assumed its present shape at the opening of January 1, the fossil record in the Cambrian rocks dates did not appear until November 12, the dinosaurs became extinct on December 26, and the first hominids appeared at 5:30 P.M. on December 31, with our historical records covering only the last forty-five seconds of the Earth Year.

Is there anything we can do that would cover a still mightier time span?

Obviously the entire universe had a beginning, with the Big Bang. The time when the Big Bang occurred can't be determined as easily or as accurately as the time when Earth and the rest of the solar system assumed its present form, and there is controversy among astronomers on the matter. However, 15 billion years is a plausible figure and it is the one, pending some good evidence to the contrary, that I generally use in my writings.

We can set 15,000,000,000 B.P. (Before the Present) as the beginning of the universe, then, and call it New Year's Minute—the midpoint of the midnight that ushers in January 1. The present moment is the midstroke of the midnight that puts an end to the following December 31. To cover the

259

entire lifetime of the universe in a single imaginary "Universe Year," each day of that imaginary year (each "Universe Day") must be 41,000,000 real years long.

As it happens, an enormous number of vitally important events that shaped the nature of the universe took place in the first few seconds after the Big Bang—even in the first few microseconds after the Big Bang. As a result, one would inevitably miss a great deal if one tried to describe everything in a year measured in the ordinary arithmetical way. What is really needed is a logarithmic scale, and I did that sort of thing in "The Crucial Asymmetry" in *Counting the Eons* (Doubleday, 1983).

Nevertheless I shall stick to an ordinary arithmetical scale for the Universe Year, as I did in the various "years" of the preceding chapter and show what I can in that way. (I'm continuing the numbering of tables from where I left off in the preceding chapter.)

9. *The Universe Year*

The Big Bang	January 1, 12:00 A.M.
Subatomic particles form	January 1, 12:00:13 A.M.
Hydrogen and helium atoms form	January 1, 12:10 A.M.
Atoms form galaxy-sized gas clouds	January 3, 10 A.M.
The Milky Way galaxy forms	February 18
The solar system forms	September 9
Life begins on Earth	October 6
First land life on Earth	December 2
First hominids appear	December 31, 9:40 P.M.
History begins	December 31, 11:59:50

As you see, the universe went through the first eighth of its history without our galaxy, and perhaps without any galaxies. (That depends, incidentally, on which of the current versions of the Big Bang are accurate. Some recent ones postulate an "inflationary universe" in which, after the Big Bang, there is a sudden, incredibly fast expansion and that may mean that galaxies may have existed almost from the start. Unfortunately, I'm not sure. I have not yet managed to grasp the inflationary universe.)

In any case, there is no doubt that the universe existed for a long time, probably seven tenths of its existence, without our solar system.

If it is true as some maintain (but I cannot make myself believe) that Earth life is the only life in the universe, then the universe went through three fourths of its existence as a vast sterility, free even from the simplest life. (How can that be credible?)

What surprises me most, though, is that the vast duration does not reduce something as petty as human history to immeasurability. Not at all! The period during which human beings have been writing chronicles of one sort or another actually occupies about ten Universe Seconds. (The last ten, of course.)

It might seem that by considering the life of the universe I have run out of useful tables. What can I call on that is still longer and grander than the total life of the total universe?

But then, length isn't all. We can find usefulness in other directions. For instance—

The Sun, with its family of planets, travels steadily about the center of the Milky Way galaxy in a nearly circular orbit, and completes one revolution in about 200,000,000 years.

Suppose we assume that the Sun's orbit has been stable; that it has not been seriously affected by stellar perturbations over its lifetime. We have no real evidence for this assumption, but there is no reason to suppose that the orbit has undergone serious changes at any time, either. And if there is no evidence either way it makes good sense to take the simplest reasonable assumption, and we'll opt for stability.

In that case it would mean that in the 4,600,000,000-year history of the solar system there has been time for the Sun and planets to have circled the galactic center 23 times.

Next, let's imagine some observer at a fixed point in the galaxy (relative to its center), one from which he saw the Sun ignite and begin to shine just as it passed him. What would he see if he remained there and studied Earth each time it returned after an interval of 200,000,000 years?

If we squeeze the lifetime of the solar system into a single "Solar System Year," then each orbit of the solar system about the galactic center would take 15.87 Solar System Days, and each of those days would represent 548,000 real years. We might prepare a table that numbers Earth's formation as 0, and then numbers each return along its orbital path from 1 to 23. The result would be as follows:

10. The Solar System Year

0—January 1   Earth assumes its present form
1—January 16   Chemical evolution
2—February 1   Chemical evolution
3—February 17   Chemical evolution
4—March 3   Chemical evolution
5—March 19   Chemical evolution
6—April 4   Bacteria (prokaryotes) appear
7—April 20   Bacteria
8—May 5   Bacteria
9—May 21   Blue-green algae (prokaryotes) appear
10—June 6   Bacteria and blue-green algae
11—June 22   Bacteria and blue-green algae
12—July 8   Bacteria and blue-green algae
13—July 24   Multicellular prokaryotes appear

THE DIFFERENT YEARS OF THE UNIVERSE

14—August 9	Multicellular prokaryotes
15—August 25	Multicellular prokaryotes
16—September 9	Multicellular prokaryotes
17—September 25	Multicellular prokaryotes
18—October 11	Eukaryotic cells develop
19—October 27	Multicellular eukaryotes (plants and animals)
20—November 12	Shellfish. Beginning of rich fossil record
21—November 28	Land life appears
22—December 14	Dinosaurs appear
23—December 31	*Homo sapiens* dominates Earth

Let me explain some points briefly. By "chemical evolution" I mean the gradual buildup of complex molecules from simple ones at the expense of various sources of energy such as solar ultraviolet, lightning, and the Earth's internal heat.

Prokaryotes (which I briefly mentioned in the previous chapter are simple cells that are considerably smaller than those of our bodies, and that lack internal complexity. They lack a nucleus, for instance, and their genetic equipment is distributed through the cell generally. The prokaryotes that still flourish today are bacteria and blue-green algae. The two are very much alike except that the blue-greens (which are not really algae, by the way) can photosynthesize and bacteria cannot.

Eukaryotes are much larger cells, with considerable internal organization, including (in particular) a nucleus. "Eukaryote" is from the Greek and means "good nucleus," while "prokaryote" means "before the nucleus." Protozoa and true algae are single eukaryotic cells, animal and plant respectively. All multicellular organisms on Earth today (including ourselves, of course) are made up of eukaryotic cells.

Multicellular prokaryotes are little more than bacterial colonies, and were a dead end. If bacteria and blue-greens survive handily today, despite the competition, it is because they occupy all sorts of niches that nothing else can or will, and because they are so incredibly fecund.

By looking at the Earth at fixed intervals, you can get a good idea of the accelerating rate of evolution. During the first five turns about the galactic center, Earth was lifeless. During the next twelve turns, it carried nothing more advanced than prokaryotic cells.

It was not till the completion of the eighteenth turn, by which time over three fourths of Earth's present age had been reached, that eukaryotic cells were developed.

But then things speed up. By the next turn we gained the potential of a good fossil record to help us, thanks to the appearance of complex multicellular organisms with parts that fossilize easily. Another turn and the land is colonized. Still another, and the dinosaurs appear.

And then the whole dramatic tale of the rise and fall of the dinosaurs, the rise of the mammals, and the coming of the hominids and modern man is all squeezed into the most recent turn of the solar system about the galactic center.

We can only wonder what there will be to see on the next turn, 200,000,000 years hence.

As long as we're considering Earth's evolution from the universe's standpoint, with talk of the Big Bang and of galactic revolutions, let's justify the title of this essay by abandoning Earth altogether now, and considering the evolution of stars—the Sun in particular—instead of that of terrestrial life.

Nearly five billion years ago the solar system existed as a huge cloud of dust and gas, a cloud that may have been there ever since the galaxy had formed, billions of years earlier still. Some impulse—a nearby supernova explosion, perhaps—

set the solar system gas cloud into contraction. Its gravitational intensity increased as a result, and the contraction was further hastened. Finally, after ten or twenty million years, the center of the cloud had contracted to a density and temperature that were sufficient to ignite hydrogen fusion. The center of the condensation "caught fire" and became a star, even while in the outer regions smaller and, therefore, cold-surface bodies (the planets) were forming.

After that, the Sun maintained its energy output by steadily fusing the hydrogen that made up by far the bulk of its contents into the somewhat more complex helium. The helium, denser than hydrogen, gathered at the solar center and this helium core grew larger and larger as ever more helium was formed and trickled down to join it.

As the helium core grew more massive, its own gravitational intensity made it condense into greater density and temperature. By the time the Sun uses up about ten percent of its total original hydrogen—something that won't happen for several billion years yet—the helium core will have grown dense enough and hot enough for helium fusion into carbon to take place.

Between the time that hydrogen fusion was initiated and the time helium fusion was, the radiational output of the Sun was reasonably constant, as it would be with any such star. During this period of time the Sun, or any star, is said to remain on the "main sequence."

In the case of the Sun, it is estimated that it will stay on the main sequence for 10 billion years altogether.

Once helium burning starts, the helium core heats up tremendously and expands. It also heats the outer hydrogen envelope, which also expands. The Sun grows larger and larger and its expanding outermost surface gradually cools to mere red heat, though the expanding surface gives it a steadily increasing *total* heat despite the cooling of the parts.

The Sun would reach its maximum volume as a "red giant" perhaps 1.5 billion years after helium burning had started, so that its total lifetime from ignition to red giant would be 11.5 billion years. (Naturally, the Sun will continue to exist and evolve after it has become a full-grown red giant, but in this chapter we won't go further.)

Other stars go through the same changes, but not necessarily at the same rate. Stars more massive than the Sun do everything more quickly. Being more massive, they have a more intense gravitational field and contract more quickly, grow denser and hotter faster, and reach ignition sooner. After ignition they fuse their hydrogen more quickly, and reach the red-giant stage more quickly too, and, for that matter, become a larger red giant. To be sure, the more massive a star, the more hydrogen it contains for fusing, but the rate of fusion goes up considerably faster than the mass of the star, so the larger the star, the shorter its stay on the main sequence.

A star three times the mass of the Sun, for instance, will complete its contraction in perhaps 3 million years, rather than the 20 million the Sun seems to have taken. It will be on the main sequence only a quarter of a billion years and will be a full-grown red giant a few million years after that.

Suppose, then, we prepare a "Sun Year" in which the total lifetime of the Sun, from ignition to full-grown red giant, is compressed into a year. Since the Sun Year would be 11.5 billion real years long, each Sun Day would be 31,500,000 real years long. We can chart the lifetime of the more massive stars in this way.

11. The Sun Year

Star ignition	January 1, 12:00 A.M.
Most massive stars become red giants	January 1, 12:45 A.M.
Star like Beta Centauri becomes red giant	January 1, 7:30 A.M.

THE DIFFERENT YEARS OF THE UNIVERSE

Star like Achernar becomes red giant	January 3, 4:00 A.M.
Star like Sirius becomes red giant	January 16
Star like Altair becomes red giant	February 1
Star like Canopus becomes red giant	March 3
Star like Procyon becomes red giant	May 5
The Sun at its present stage	May 25
Helium burning begins in the Sun	November 12
The Sun becomes full-grown red giant	December 31

As you see, the Sun is still in its vigorous middle age, with not quite half of its useful life gone. Nor is there any need to worry about the fact that, inexorably, after helium burning begins, the Sun will grow steadily hotter so that life on Earth will become impossible. Indeed, when the Sun is a full-grown red giant it will expand till it is close enough to the Earth to heat it to a baked cinder. It may even engulf it altogether.

It should, however, be at least five or six billion years before the heat is really on, and it would take an incurable optimist to suppose that we won't have managed to find something altogether different as a means of doing ourselves in. We won't have to wait around for a heating Sun.

Even if we survive, then by helium-burning time we will have evolved into something unrecognizable as human (though, we can always dimly hope, something better than human).

If we or a successor species exist when the Sun is at helium ignition, it is inconceivable that our technological level will not have reached the point where we can leave Earth easily and retreat to the outer solar system where the Sun's newly enormous total heat will be beneficent rather than otherwise. In fact we can be certain that, long before the Sun's heat becomes a problem, humanity or its descendants will have transferred the scenes of its activity to planets circling other, younger stars, or to independent artificial worlds.

It might occur to you, by the way, that if a star like Beta Centauri works its way through the main sequence in a mere five and a half Sun Hours and is gone, so to speak, before sunrise on the first day of the year, how can it be that Beta Centauri is shining serenely in the skies of the southern hemisphere right now?

Ah, but Table 11 is based on the supposition that a whole group of stars of various masses (but all more massive than the Sun) was ignited at the same time. This is not the case with the real stars in the galaxy about us. Beta Centauri has a total life on the main sequence of not more than 10 million years, yet it shines in the sky now because it was formed *less* than 10 million years ago.

All stars more massive than the Sun are relative newcomers to the scene—otherwise all would have gone red giant and been in a state of collapse by now. Many spiral galaxies (including the Milky Way) are still littered with clouds of dust and gas and these can, under the proper conditions, condense into whole crowds of stars. There are small, intensely dark patches called "Bok globules" after the astronomer Bart J. Bok, who first called attention to them, and these may be stars in the actual process of forming as we watch.

Just as there are stars that are more massive than the Sun, and that are therefore larger, more luminous, hotter, and shorter-lived, there are also stars that are less massive than the Sun and therefore smaller, less luminous, cooler, and longer-lived.

The small stars make no great splash in the sky; we are much more aware of the large, bright ones. However, in the case of stars, as in the case of almost any large group of similar substances, be they galaxies, pebbles, or insects, the smaller ones are more numerous than the larger ones. For every star as massive as, or more massive than, the Sun, there are six or seven stars that are less massive than the Sun.

The smallest of the stars are cool enough to be only red-hot. Unlike the red giants, the small stars do not have great size to make up for the dimness of the parts. The small stars are therefore dim altogether—so dim that although they may be quite close to us, they can, even so, only be seen by telescope.

These small stars are called red dwarfs, and they are so stingy with their energy that they last a surprisingly long time. A very small red dwarf, one just large enough to sustain a feeble nuclear fusion, can make its relatively small fuel supply last through 200 billion years on the main sequence. This means that no red dwarf has ever left the main sequence. The universe simply isn't old enough to have worn one of them out.

Let us then set up a "Red Dwarf Year," by which I mean 200,000,000 years compressed into a single year (which gives us something longer than the present lifetime of the universe—much longer), and see what the stars look like from that standpoint. Each Red Dwarf Day would, by this system, be 548,000,000 years long.

12. The Red Dwarf Year

Star ignition	January 1, 12:00 A.M.
Star like Sirius becomes a red giant	January 1, 10 P.M.
Star like Altair becomes a red giant	January 2, 8 P.M.
Star like Canopus becomes a red giant	January 3, 3 P.M.
Star like Procyon becomes a red giant	January 7, 7 A.M.
Star like the Sun becomes a red giant	January 21
Star like Alpha Centauri B becomes a red giant	February 24
Star like Alpha Centauri C becomes a red giant	December 31

If we could imagine a red dwarf's having consciousness and watching the universe, it might take note, rather sardonically, of all the big firecrackers that come and go in rapid flashes, while it and its fellow red dwarfs burn steadily onward in their dim, quiet way.

To be sure, new firecrackers would arise, but it is quite likely that the red dwarfs would continue to shine past them also. In fact, when the gas and dust of those various galaxies that have such clouds (many galaxies are dust-free) are consumed, and the bright stars have all gone past the red-giant stage and collapsed into dimness, then the universe will flicker feebly in the light of the only normal stars left, the red dwarfs.

But eventually, if the universe is open and is expanding forever, the last red dwarf will blink out, too, and there will be no main sequence stars left at all.